J.G. White

Starling Facts

J.G. White

Starling Facts

ISBN/EAN: 9783337038366

Printed in Europe, USA, Canada, Australia, Japan

Cover: Foto ©berggeist007 / pixelio.de

More available books at **www.hansebooks.com**

STARTLING FACTS:

OR,

DEEDS OF DARKNESS DISCLOSED

RELATIVE TO

AURICULAR CONFESSION,

AND ITS RELATIONS TO

SACERDOTAL CELIBACY, CONVENTS, MONASTERIES, MORALITY, AND CIVIL AND RELIGIOUS LIBERTY.

BY

REV. J. G. WHITE,

AUTHOR OF THE "PROTESTANT MISSIONARY," AND OTHER ANTI-ROMAN PUBLICATIONS.

"And have no fellowship with the unfruitful works of darkness, but rather reprove them."—Paul.

CINCINNATI:
PUBLISHED BY THE AUTHOR.
1875.

PREFACE.

THE present volume is offered to the public as the first of a series of works in which the author proposes to expose Romanism and defend Protestantism. This work devotes special attention to *Auricular Confession,* its corrupting and intolerant influences. The confessional is regarded as the main pillar of Popery, the instrument of a despotic clerical power, and the arch-key of the whole superstructure of the Papacy, without which it would crumble to the dust.

This book is intended to be an embodiment of facts and documentary evidence of the pernicious influences of the confessional. It is a beacon-light to warn Protestants against the seductive influence of the confessional in connection with professed sacerdotal celibacy and convent life. It is also intended to show that Auricular Confession degrades and enslaves its votaries, and that through it the Roman clergy are endeavoring to subvert and destroy the principles of civil and religious liberty throughout the world; and that their energies are especially concentrated against the Government of the United States, with a determination to destroy it, and on its ruins establish a Papal despotism.

This book was written in the midst of numerous and pressing professional engagements; and if it does not possess variety, it should not be attributed to the monotony of surrounding circumstances. It has been written while traveling thousands of miles, and at intervals between lectures, at hotels, between sermons, at protracted meetings, when often surrounded by the domestic circle, and occasionally at home, when resting from the fatigues of a journey, and at all hours of day and night. The numerous references to authorities has required the continued presence of a small select traveling library of Roman books. We submit the work to the careful consideration of a generous American people, conscious that, while its style may be subject to criticism, it contains important facts which challenge investigation.

Frequent abortive efforts have been made by Romanists to assassinate the author, and he has the positive evidence that they intend to take his life. He therefore puts these facts in form to speak for themselves; and, if he falls by the hand of an assassin, he will fall in defense of outraged and insulted virtue, and fall *fearlessly* at his post, battling for virtue, for liberty, for truth, for the right, for the salvation of man, and for the honor of God.

THE AUTHOR.

JACKSONVILLE, ILL., *Feb.* 1, 1875.

CONTENTS.

CHAPTER I.
INTRODUCTORY, ... 7

CHAPTER II.
AURICULAR CONFESSION DEFINED, ... 12

CHAPTER III.
AURICULAR CONFESSION FURTHER DEFINED, . . . 26

CHAPTER IV.
THE SEAL OF SACRAMENTAL CONFESSION, ... 33

CHAPTER V.
THE CONFESSIONAL, ... 52

CHAPTER VI.
SINS, MORTAL AND VENIAL, ... 63

CHAPTER VII.
POWER OF THE KEYS, ... 76

CHAPTER VIII.
THE CLERGY AND CONCUBINES, ... 90

CHAPTER IX.
Clerical Seduction, how Concealed, ... 99

CHAPTER X.
Corruption of the Confessional, ... 109

CHAPTER XI.
Corruption of the Confessional, Continued, . . 115

CHAPTER XII.
Corruption of the Confessional, Continued, . . 134

CHAPTER XIII.
The Confessional a Thief-trap, ... 143

CHAPTER XIV.
The Confessional enslaves Men, ... 149

CHAPTER XV.
Protestant Slaves to the Confessional, . . . 156

CHAPTER XVI.
Prison-pens for American Daughters, ... 170

CHAPTER XVII.
Papal Conspiracy aided through the Confessional, . 174

CHAPTER XVIII.
Romish Intolerance enforced through the Confessional, 209

AURICULAR CONFESSION EXPOSED.

CHAPTER I.

INTRODUCTORY.

THE intolerant, despotic power of the Church of Rome over the souls and bodies of men is maintained by the direct influence of the confessional. The crime and licentiousness of cities and nations has been, and is now, in proportion to the unrestrained influence of the confessional. Civil and religious liberty struggle in vain for existence where its obligations are universally recognized. It therefore becomes the duty of all true patriots to investigate the principles of an institution, the influence of which is evil, and only evil, continually.

To comprehend the moral degradation and abject servitude which result from the Romish Confessional, it is necessary, first, to understand its principles, its obligations, its practices, and its legitimate results. Such is the nature of the subject, and such are the facts connected with its investigation, that a regard for decency precludes the possibility of full disclosures.

To form a correct estimate of the horrible corruption of the confessional, reference must be had to the secret theology and ritual of the Roman clergy, much of which should not be translated, nor published for promiscuous readers. In this work we can only coast along the shore of a boundless ocean of filth. We dare not disturb the scum of its smallest adjacent cesspool. Its exhalations are infected with moral pestilence; and protracted contact with its poisoned waters often results in eternal death.

The necessity for Auricular Confession is predicated on the false and blasphemous assumption of the Roman clergy, who arrogate to themselves the titles of vicegerents and vicars of Jesus Christ, possessed of judicial power as God to forgive or retain sin, and to save or damn the souls of men at pleasure. And so absolute is this power that if a priest, in confession, refuse to pardon a penitent, Jesus Christ himself can not do it.

Notwithstanding this blasphemous assumption of power, priests are compelled to admit that they do not possess all the attributes of God; they are not omniscient nor omnipresent; and they are chiefly dependent on the extorted confessions of transgressors for their knowledge of sins committed. Under these circumstances they are as liable to be mistaken, deceived, and imposed upon, as other men. When confession is made, the priest does not know whether it is true or false, partial or thorough, feigned or sincere. And the penitent, if sincere, does not know whether he has confessed all, or forgotten a part of his sins. And if the penitent is sincere, and is sure that he has confessed all *mortal sins*, and the priest has pronounced absolution in the usual form, the penitent does not know that the priest had the requisite *intention*, without which his pretended absolution is a blasphemous ecclesiastical farce.

In order to expedite this difficult work of Romish pardon and salvation, the clergy have instituted Auricular Confession, which will receive attention in the following pages. This book is intended to meet the wants of the general reader, which fact will preclude the possibility of extended quotations from the corrupt, secret, Latin theology of the Roman clergy. The most chaste extracts are only admissible in consideration of correcting or preventing the evil influence. Ministers and men of age are referred to the original, which are before us, and can not be successfully denied nor defended.

Auricular Confession, in the hands of the Roman clergy, is the masterpiece of the devil's workmanship, the arch-key of the whole superstructure of clerical power. Strike down the confessional, and professed sacerdotal celibacy will be discarded, convent life will lose many of its attractions, " foundling institutions " will be less patronized, and " Magdalene institutions " and houses of the " *Good Shepherd*," will be less in demand for clerical prostitutes. Abolish the confessional, and the clerical power of Rome will vanish with it, and millions who are now crushed by Popish despotism, will enjoy civil and religious liberty. No class of men understand these facts better than the Roman clergy; and hence their *fury* when these secret abominations are exposed.

We have had much experience on this subject. Repeated mob violence, and efforts at assassination, have been employed in vain to suppress the facts. The truth must and shall be proclaimed. The abominations of the confessional shall be exposed till its corrupting influences are understood, and until it shall be declared a nuisance, and suppressed by legal enactments. If there is a law in the land for the suppression of brothels, it might, with equal propriety, be enforced also against this prolific source of licentiousness.

Let Protestant parents understand the relation which priests sustain to nuns in the confessional, and they will cease to patronize convents. Let husbands understand the libidinous questions which bachelor priests are authorized to propound to their wives in confession, and their just indignation will demand redress. Let the people understand that the Roman clergy are the truculent minions of an ecclesiastical despot, and that through the confessional they are prostituting virtue, corrupting society, and endeavoring to subvert the institutions of the nation, and enlightened public sentiment will consign the confessional to merited infamy.

We are impelled to the publication of this work by the fact that Protestants generally have no just conception of the "mystery of iniquity" now practiced in our midst by the "mother of harlots." Also, from the fact that common decency will forever preclude the possibility of disclosing the worst to a select company of men, much less in a work for general readers.

The well-being of society demands that sufficient light be shed on this most detestable system of darkness to guard the unsuspecting against its seductive influences.

This work is intended to arrest attention, and disclose such facts as may be prudently presented to the general reader, with the hope that men of mature years, and especially ministers of the Gospel, may be induced to examine more thoroughly this prolific source of licentiousness, which is subverting the virtue of youth, and jeopardizing the souls of millions. We predicate our statements on books and documents before us which challenge investigation, and we defiantly hurl the facts in the face of the Roman clergy. Such a system of unblushing licentiousness shall not escape merited rebuke and public exposure.

Trusting for success to the justness of our cause, and to that power which guided the sling of David, we hurl this pebble of truth at the brazen pate of the " man of sin," and pray God to smite him to the dust.

CHAPTER II.

AURICULAR CONFESSION DEFINED.

AURICULAR Confession is a modern invention, a device of wicked men, and a prolific source of crime and licentiousness. It is not authorized in the Word of God, nor sanctioned by common sense. It was not known to Moses nor the prophets, and it was not taught by Jesus Christ, nor by his apostles. It originated in ignorance and superstition, and can only be perpetuated by this influence. Auricular Confession is literally confession in the ear of a priest in order to obtain judicial absolution from all mortal sins committed after baptism. Roman theology teaches that baptism pardons original sin, and that the Roman clergy, by the "power of the keys," grant judicial pardon as God, for all mortal sins committed by their faithful after baptism, and is thus defined: "Confession, then, is defined a sacramental accusation of one's self, made to obtain pardon by virtue of the keys." (Catechism of Trent, p. 191.)

Previous to the Lateran Council, A. D. 1215, the confession of sin was an optional thing in the Church of Rome. In the midnight darkness of the world it had increasing popularity for two centuries. The flagrant licentiousness of bishops and popes of this period demanded secrecy, or otherwise the entire suspension of confession in any form. Confession to God, and public confession in presence of the Church, had been long practiced; but the debauchery of the clergy and popes, and consequent corruption of the people, brought public confession into disrepute and furnished strong inducements to conceal vice.

Confession had been recommended; but it had no sovereign sanctions to enforce it, no canon or bull to compel it throughout the Roman Church previous to A. D. 1215, and the new dogma was not confirmed till the Council of Trent in its fourteenth session, A. D. 1557, the canons of which clearly defined the doctrine, as follows:

"CANON 1. Whoever shall affirm that penance, as used in the Catholic Church, is not truly and properly a sacrament, instituted by Christ our Lord, for the benefit of the faithful, to reconcile them to God, as often as they shall fall into sin after baptism, LET HIM BE ACCURSED."

"CANON 3. Whoever shall affirm that the words of the Lord our Saviour, 'Receive ye the Holy Ghost; whose sins you shall forgive they are forgiven them, and whose sins you shall retain they are retained;' are not to be understood of the power of forgiving and retaining sins in the sacrament of penance, as the Catholic Church has always from the very first understood them; but shall restrict them to the authority of preaching the Gospel, in opposition to the institution of this sacrament, LET HIM BE ACCURSED."

"CANON 6. Whoever *shall deny that sacramental confession* was instituted by Divine command, or that it *is necessary to salvation;* or shall affirm that the practice of *secretly confessing to the priest alone,* as it has been ever observed from the beginning by the Catholic Church, and is still observed, is foreign to the institution and command of Christ, and is a human invention, LET HIM BE ACCURSED."

"CANON 7. Whoever shall affirm that, in order to obtain forgiveness of sins in the sacrament of penance, it is not by Divine command necessary to confess all and every mortal sin which occurs to the memory after due and diligent premeditation, including *secret offenses*, etc., LET HIM BE ACCURSED."

"CANON 8. Whoever shall affirm that the confession of every sin, according to the custom of the Church, is impossible and merely a human tradition, which the pious should reject; *or that all Christians, of both sexes, are not bound to observe the same once a year*, according to the constitution of the great Council of Lateran, and therefore that the faithful in Christ are to be persuaded not to confess in Lent, LET HIM BE ACCURSED."

"CANON 9. Whoever shall affirm that the priest's sacramental absolution is not a judicial act, but only a ministry, to pronounce and declare that the sins of the party confessing are forgiven, so that he believes himself to be absolved, even though the priest should not absolve seriously, but in jest; or shall affirm that the confession of the penitent is not necessary in order to obtain absolution from the priest, LET HIM BE ACCURSED."

These are only a portion of the canons which define Auricular Confession. In conformity to the above decrees, Pope Pius V. approved the Catechism of Trent as the infallible exponent of canon law, and it is now so regarded. The Bible itself is required to conform to its teaching. On page 190 we have the following, as translated into English by Rev. J. Donovan, Professor, etc., Royal College, Maynooth:

"Contrition, it is true, blots out sin; but who is ignorant, that to effect this, it must be so intense, so ardent, so vehement, as to bear a proportion to the magnitude of the crimes which it effaces? This is a degree of contrition which few reach, and hence, through perfect contrition alone very few indeed could hope to obtain the pardon of their sins. It therefore became necessary that the Almighty, in his mercy, should afford a less precarious and less difficult means of reconciliation and of salvation; and this he has done, in his admirable wisdom, by giving to his Church the keys of the kingdom of heaven. According to the doctrine of the Catholic Church, a doctrine firmly to be believed and professed by all her children, if the sinner have recourse to the tribunal of penance with a sincere sorrow for his sins, and a firm resolution of avoiding them in future, although he bring not with him that contrition which may be sufficient of itself. to obtain pardon of sin, his sins are forgiven by the minister of religion, through the power of the keys. Justly, then, do the holy fathers proclaim that by the keys of the Church the gate of heaven is thrown open; a truth which the decree of the Council of Florence, declaring that the effect of penance is absolution from sin, renders it imperative on all unhesitatingly to believe."

Again, on page 192:

"Invested, then, as they are, evidently appointed judges of the matter on which they are to pronounce; and as, according to the wise admonition of the Council of Trent, we can not form an accurate judgment on any matter, or award to crime a just proportion of punishment, without having previously examined and made ourselves acquainted with the cause; hence, arises a necessity, on the part of the penitent, of making known to the priest, through the medium of confession, each and every sin. This doctrine, a doctrine defined by the holy Synod of Trent, the uniform doctrine of the Catholic Church, the pastor will teach. . . . When, with uncovered head and bended knees, with eyes fixed on the earth, and hands raised in supplication to heaven, and with other indications of Christian humility not essential to the sacrament, we confess our sins; our minds are thus deeply impressed with the clear conviction of the heavenly virtue of the sacraments, and also of the necessity of humbly imploring and earnestly importuning the mercy of God. . . . To obtain admittance into any place, the concurrence of him to whom the keys have been committed is necessary; and therefore, as the metaphor implies, to gain admission into heaven, its gates must be opened to us by the power of the keys, confided by Almighty God to the care of his Church.

"This power should otherwise be nugatory: if heaven can be entered without the power of the keys, in vain shall they to whose fidelity they have been intrusted, assume the prerogative of prohibiting indiscriminate entrance within its portals."

Again, on page 193:

"According to the Council of Lateran, which begins '*Omnis utriusque sexus*,' no person is bound by the law of confession until he has arrived at the use of reason, a time determinable by no fixed number of years. It may, however, be laid down as a general principle, that children are bound to go to confession as soon as they are able to discern good from evil, and are capable of malice; for when arrived at an age to attend to the work of salvation, every one is bound to have recourse to the tribunal of penance, without which the sinner can not hope for salvation. In the same canon the Church has defined the period, within which we are bound to discharge the duty of confession: it commands all the faithful to confess their sins at least once a year. If, however, we consult for our eternal interests, we will certainly not neglect to have recourse to confession as often, at least, as we are in danger of death, or undertake to perform any act incompatible with the state of sin, such as to administer or receive the sacraments."

Again, on page 194:

"All mortal sins must be revealed to the minister of religion; venial sins, which do not separate us from the grace of God, and into which we frequently fall, although, as the experience of the pious proves, proper and profitable to be confessed, may be omitted without sin, and expiated by a variety of other means. Mortal sins, as we have already said, although buried in the darkest secrecy, and also sins of desire only, such as are forbidden by the Ninth and Tenth Commandments, are all and each of them to be made a matter of confession. Such secret sins often inflict deeper wounds on the soul than those which are committed openly and publicly. . . . Some circumstances are such as, of themselves, to constitute mortal guilt; on no account or occasion whatever, therefore, are such circumstances to be omitted. Has any one imbrued his hands in the blood of his fellow-man? He must state whether his victim was a layman or an ecclesiastic. Has he had criminal intercourse with any one? He must state whether the female was married or unmarried, a relative, or a person consecrated to God by vow."

THE PRIEST FORGIVES ALL SORTS OF SINS.

Again, on page 196 :

"But in case of imminent danger of death, when recourse can not be had to the proper priest, that none may perish, the Council of Trent teaches, that, according to the ancient practice of the Church of God, it is then lawful for any priest not only to remit all sorts of sins, whatever faculties they might otherwise require, but also to absolve from excommunication."

These extracts from canons and the Catechism of the Council of Trent may be examined with care, as we shall presently have use for them. It may also be observed that while the priest may be in flagrant violation of the *Seventh* Commandment of the decalogue—Sixth of the Douay Bible—he professes to have power to absolve his accomplice in crime from all other sins, and in case of danger of death from *that sin* also. This last fact will receive attention in subsequent pages.

All the approved theology of the Roman Church, and all public and private instruction, is required to conform strictly to the teaching of the Council of Trent. This fact will be more apparent by reference to the catechisms and manuals in the hands of the laity and the secret theology for the instruction of the clergy.

We here introduce extracts from the common catechisms and other books in general use throughout the United States, with the approbation of the bishops.

POOR MAN'S CATECHISM.

On page 140, we find the following:

"THE THIRD PRECEPT OF THE CHURCH EXPOUNDED.

" *Q.* What is the third precept of the Church?

" *A.* To confess our sins to our pastor at least once a year.

" *Q.* Why was this commanded?

" *A.* Because libertines would not otherwise have done it once in many years.

"INSTRUCTION.—This precept is contained in a canon of the fourth Council of Lateran, under Innocent the Third, held in the year of our Lord 1215, which was confirmed by the Council of Trent, Sess. xiv., c. v., and can. 8, whereby all the faithful, of both sexes, are strictly enjoined to confess their sins to their proper pastor once in a year at least; and to receive the sacrament of the holy Eucharist at Easter, as soon as they come to years of discretion sufficient for each sacrament. This precept, then, begins to bind us as soon as we begin to have the full use of reason, so as to commit mortal sin, and to be capable of the sacrament, which, in some, is sooner, in some later. The Church does not particularly prescribe the time of the year when we ought to confess; yet, as we are obliged to communicate at Easter, which can not be rightly done in a state of sin, it is evident that all those who, at that time, are in mortal sin are obliged then to confess.

"Though the precept of the Church obliges us to confess but once a year to restrain libertines, yet many circumstances may occur, in which, by the divine precept, we are obliged to confess oftener. 1. In all dangers of life, as when dangerously sick, or condemned to die, or when soldiers are to go to battle, or merchants to go a hazardous voyage, and are conscious of any mortal sin to themselves; in such dangers (life so uncertain), they are bound to confession; because, in all perils of life, we are bound to prepare ourselves for death. Ought any one that knows himself to be in a bad state, considering the uncertainty of life, run the risk of a delay? 2. Before we receive the other sacraments, if guilty of mortal sin, we are bound, first, to confess; because such sin is opposite to divine grace, and must, of necessity, hinder the blessed effect of the sacraments we receive, baptism excepted; for baptism being the first sacrament, by it we must be made Christians before we can receive any of the *Christian sacraments;* therefore, sacramental confession is not required before baptism, but only contrition in adult persons. Neither does every sort of confession satisfy our obligation; but we are to make a true and entire confession, which can not be done without a previous and careful examen of our life and conscience."

Bishop Butler's Catechism, which is approved and in use on both continents, contains the following, on page 41 :

" *Q.* What means the commandment of confessing our sins at least once a year?

" *A.* It means that we are threatened with very severe penalties by the Church if we do not go to confession within the year.

" *Q.* Does a bad confession satisfy the obligation of confessing our sins once a year?

" *A.* So far from it that it renders us more guilty by the additional crime of sacrilege.

" *Q.* Is it sufficient to go but once a year to confession?

" *A.* No; frequent confession is necessary for all those who fall into mortal sin, or who desire to advance in virtue.

" *Q.* At what age are children obliged to go to confession?

" *A.* As soon as they are capable of committing sin; that is, when they come to the use of reason, which is generally supposed to be about the age of seven years."

Thus the obligation binds all, beginning with children seven years of age. And to this agrees the Catechism of Trent, and numerous other catechisms published by bishops on both continents.

A general catechism for the use of Romanists in the United States, and "approved by the Most Rev. John

Hughes, D. D., Archbishop of New York," contains, on page 41, the following:

"*Q.* What is confession?

"*A.* Confession is to accuse ourselves of all our sins to a priest, in order to obtain absolution.

"*Q.* How must we declare our sins?

"*A.* We must declare the number of our sins, and their different kinds.

"*Q.* Must we declare all our sins?

"*A.* We must declare all our mortal sins; for, if we were to conceal willfully any mortal sin, we should not obtain the forgiveness of any, and we should besides commit a sacrilege.

"*Q.* What must we do to obtain an exact knowledge of all our sins?

"*A.* We must carefully examine our consciences upon the commandments of God and of the Church, and see in what we have sinned upon each of these commandments.

"*Q.* In what sentiments should we present ourselves before the priest, when we are going to confession?

"*A.* We should kneel, and begin our confession as criminals who implore the mercy of God, viewing Jesus Christ in the person of the priest."

Again, page 43:

"*Q.* How ought we to accuse ourselves of our sins?

"*A.* We ought to accuse ourselves of our sins with much sincerity and humility, and begin by those we have most difficulty in declaring.

"*Q.* What should we do when the confessor puts us questions?

"*A.* We should answer the questions of the confessor clearly and with simplicity

"*Q.* What should we do when we have finished declaring our faults?

"*A.* After telling our sins, we should finish the Confiteor, *I confess to Almighty God*, etc.; then listen with attention to the advice which the confessor may think proper to give.

"*Q.* What ought we to do whilst the priest is giving absolution?

"*A*. Whilst the priest is giving the absolution, we ought humbly to bow down our heads and renew our act of contrition with all the fervor we are capable of.

"*Q*. What is absolution?

"*A*. Absolution is the forgiveness of our sins, which the priest imparts in virtue of the power he has received from Christ.

"*Q*. Can all priests exercise this power?

"*A*. Only such priests as are approved of by the bishop can give absolution."

Bishop David's Catechism has had a wide circulation in Kentucky and adjacent States nearly forty years, and the late edition contains the following—pages 103–105:

"LESSON XIX.

"OF CONFESSION.

"*Q*. What is confession?

"*A*. Confession is the declaring of all our sins to a priest duly authorized, in order to receive absolution.

"*Q*. Is confession necessary to obtain the forgiveness of our sins?

"*A*. Yes; confession is necessary to obtain the forgiveness of all mortal sins committed after baptism.

"*Q*. When did our Savior command it?

"*A*. Our Savior enjoined the confession of sins, when he gave to his apostles the power of forgiving and of retaining them.

"*Q*. How do you show this?

"*A*. Because they could not know what sins to forgive, and what sins to retain, if they were not declared to them.

"*Q*. Has confession been the constant practice of the Church in all ages?

"*A*. Yes; the faithful of all ages have had recourse to confession, to obtain the forgiveness of their sins.

"*Q*. Can nothing excuse from that obligation?

"*A*. Nothing, but impossibility, can excuse from confessing mortal sins.

"*Q*. What must be the qualities of our confession?

"*A.* Confession, to be good, must have these three qualities: it must be humble, sincere, and entire.

"*Q.* How is it *humble?*

"*A.* We must declare our sins with sorrow and repentance, seeking no excuse.

"*Q.* How is it *sincere?*

"*A.* We must declare our sins as we know them, without making them greater or lesser than they are.

"*Q.* How is it *entire?*

"*A.* We must declare at least all our mortal sins which we remember, after a due examination.

"*Q.* Is it enough to tell the different kinds of sins we have committed?

"*A.* No; we must also tell the number as nearly as we can, and the chief circumstances that may increase our guilt.

"*Q.* What if one would have no sorrow for his sins, or would conceal a mortal sin in confession?

"*A.* He would, in both cases, commit a great crime, by telling a lie to the Holy Ghost, and make his confession invalid and worth nothing.

"*Q.* What must be done by those who, either from negligent examination, or by concealing their sins, or for want of contrition, have made an invalid confession?

"*A.* They must,—

"1. Make over again that confession, and all those which have followed it.

"2. Accuse themselves of all the sacrilegious receptions of communion and other sacraments.

"3. Do penance for them.

"*Q.* Are there not occasions in which a general confession is necessary?

"*A.* A general confession is necessary for those who never yet approached the sacrament of penance with the necessary disposition, or have a reasonable doubt whether they ever did.

"*Q.* What must one do who feels ashamed to declare some sin in confession?

"*A.* He must,—

"1. Earnestly beg of God the grace of surmounting that false shame.

"2. Look upon the pain of confessing his sin as the first penance to be done for it.

"3. Make use of considerations that may help him to overcome it.

"*Q.* What are these considerations?

"*A.* 1. That the priest is the father and friend of his soul.

"2. That the priest is bound, under pain of damnation, to absolute secrecy and silence with regard to the sins he hears in confession.

"3. That the sins one would thus conceal from one man will be revealed by Jesus Christ, at the last day, to the whole world."

Again, from St. Liguori:

"He who has offended God by mortal sin has no other remedy for his damnation but the confession of his sin. *But, if I am sorry for it from my heart? If I do penance for it during my whole life? If I go into the desert, and live on wild herbs, and sleep on the ground?* You may do as much as you please; but if you do not confess every mortal sin which you remember, you can not obtain pardon. I have said, *a sin which you remember;* for, should you have inculpably forgotten a sin, it has been pardoned indirectly, if you had a general sorrow for all your offenses against God. It is sufficient for you to confess it whenever you remember it. But if you have concealed it voluntarily, you must then confess not only the sin which has been concealed, but also the others which have been confessed; for the confession was null and sacrilegious.

"Accursed shame! How many poor souls does this shame send to hell! St. Teresa used to say to preachers, 'Preach, dearly beloved priests, preach against bad confessions; for it is on account of bad confessions that the greater part of Christians are damned.'" (St. Liguori on the Commandments and Sacraments, p. 219.)

Again:

"A penitent at confession should imagine himself to be a criminal condemned to death, bound by as many chains as he has sins to confess, and presenting himself before a confessor, who holds the place of God, and who alone can loose his bonds and deliver him from hell. Hence, he must speak to the confessor with great humility." (p. 227.)

Again:

"Let all who choose to advance in the way of God obey their confessor, who holds the place of God; he who acts thus is certain that he shall not have to render to God an account of the actions which he performed through obedience." (p. 196.)

"OF THE MANNER OF CONFESSING OUR SINS.

"*Q.* What must we do, when kneeling at the feet of the priest?

"*A.* We must behold in the person of the priest that of Jesus Christ himself, in whose name he sits there.

"*Q.* How must we consider ourselves?

"*A.* Each penitent must consider himself a criminal, who appears before his judge.

"*Q.* How must we begin our confession?

"*A.* We must,—

"1. Make the sign of the cross.

"2. Ask the priest's blessing, saying, *Bless me, father! for I have sinned.*

"3. Say the Confiteor, or, *I confess to Almighty God,* etc., as far as, *Through my fault,* etc.

"*Q.* What are we to do next?

"*A.* We must tell,—

"1. The time of our last confession.

"2. Whether we received absolution.

"3 Whether we have performed the penance enjoined upon us.

"4. Whether we did not forget any thing in our last confession.

"*Q.* What is the best form of confession?

"*A.* It is to say, *I accuse myself of such and such a sin, so many times.*

"*Q.* By what sin is it proper to begin our confession?

"*A.* It is proper to begin our confession by the sin which gives us most uneasiness." (Bishop David's Catechism, p. 107.)

This obligation of confession is enforced throughout the Roman Church. It is taught in all works on theology, and in all catechetical instruction. In the Roman system of theology, Auricular Confession is an indispensable condition of reconciliation and salvation. A neglect of the confessional deprives the neglector of the right to the ordinances and immunities of the Church, and forever excludes him from heaven. The ingenuity of the Church has been taxed to impress these sentiments, and compel the faithful observance of the confessional. The more effectually to accomplish this work, penitents are continually threatened with endless perdition if they die in the neglect of it.

The following from Dr. Butler's Small Catechism, page 27, expresses the approved doctrine of the Church:

"*Q.* Are any other condemned to hell beside the devils or bad angels?

"*A.* Yes; all who die enemies to God; that is, all who die in the state of mortal sin.

"*Q.* Can any one come out of hell?

"*A.* No; *out of hell there is no redemption.*"

The only exception is, where they say the Virgin Mary sometimes interposes, and rescues souls doomed to endless perdition. Priests, for money, pretend to deliver souls from purgatory—a place which does not exist; but it is reserved for Mary only to deliver from hell. (Glories of Mary, p. 123, etc.)

CHAPTER III.

CONFESSION FURTHER DEFINED.

THE above extracts disclose many startling facts worthy of careful consideration, some of which we here enumerate:

1. The penitent, when kneeling at the feet of the priest "must behold in the *person* of the *priest* that of Jesus Christ himself, in whose name he sits there." This certainly requires powerful organs of vision or a more powerful imagination to see in the person of any priest, whether drunk or sober, "*that of Jesus Christ himself.*" What strange perceptions must Romanists have when they can at one time behold Jesus Christ in the person of a bloated, licentious priest, and at another time in a drop of wine, or in the small dust of a wafer.

2. "Each penitent must consider himself a *criminal* who appears before his *judge*." How degrading this abject servitude; how lost to self-respect must be the victims of superstition who can voluntarily surrender soul and body to the dictation and domination of arrogant, self-constituted despots.

3. What a blasphemous assumption for mortal man, inflated with a self-righteous pomposity, and with his shirt-collar the back side before, in a long gown or petti-

coat, or with his coat-tail a foot longer than other men's, strutting in sanctimonious hypocrisy, proclaiming himself God, with power over three worlds—heaven, earth, and hell—and ability to save or damn the souls of men at pleasure; pretending to open and shut heaven and hell at discretion, to grant judicial pardon as God, when (unless he is better than required by his theology, and better than many of his order) he is living in debauchery, indulging his appetites, passions, and propensities. Shame on such blasphemous, heaven-daring, hell-deserving insolence, which, in hypocritical mask, and in virtue's guise, attempts to

"Steal the livery of the court of heaven,
To serve the devil in;
And transact villainies that common sinners durst not meddle with."
"Oh judgment! thou hast fled to brutish beasts,
And men have lost their reason."

4. The priest, in the confessional *as God*, pretends to forgive the sins of others, when at the same time he is liable at any moment to be eternally damned if he should disclose secrets from the confessional. Thus "the man of sin is revealed; the son of perdition who as God sitteth in the Temple of God." (2 Thess. ii.)

5. The penitent must "earnestly beg of God the grace of surmounting that *false shame*," etc. Thus it seems that in the work of seduction the "Mother of Harlots" is not limited within the ordinary limits of sensual brutality, but guided by licentious theology, her voluptuous sacerdotal seducers are authorized to instruct

their confiding penitents to "*beg of God the grace of surmounting that false shame.*" What more fiendish plot could have been devised to prostitute virtue and debase society?

Modesty, the guardian angel of virtue, must be sacrificed to gratify the avarice and lust of clerical pretenders, and the intended victims are required to "*beg God*" to assist in the soul-destroying work.

O insulted justice! how long wilt thou stay thy avenging arm, and permit the "Whore of Babylon" to revel unrestrained, "drunk with the blood of saints," and virtue by her trampled to the dust? Is there not some hidden curse, some bolt of heaven, red with uncommon wrath, to blast a system which holds fiendish carnival amidst the ruins of fallen virtue, and laughs to scorn the dying agonies of lost souls?

6. To give respectability to this seductive system of clerical debauchery, we are assured by the corrupters that "the faithful in all ages have had recourse to confession to obtain the forgiveness of their sins." By this declaration they evidently intend to teach that Auricular Confession has been practiced in all ages. This is an unmitigated falsehood, only worthy the Jesuit system of iniquity which it is intended to propagate.

7. One of the reasons assigned for Auricular Confession is the ignorance of the clergy, "because they could not know what sins to forgive and what sins to retain, if they were not declared unto them." This is an honest confession, and it is not all nor the worst of it; they are

not only ignorant of sins committed, but they have no power to forgive sins, whether known or otherwise. It is simply a blasphemous assumption, not authorized by the word of God, nor consistent with reason. The Bible throughout teaches that God only can forgive sin. God is omniscient; he knows all the thoughts, motives, desires, purposes, words, and actions of men. Not so the Roman clergy; they do not so much as know the deception of their own hearts; and they have no more power to forgive sins *judicially* than had Judas Iscariot, or Simon the sorcerer. Many of them violate both the laws of God and man, and pretend each to grant the other absolution. This looks very much like Satan casting out devils. Can it be possible that the Roman clergy are such consummate simpletons as to be deceived by their own clerical jugglery? Has their reason become stultified, or have they been given over to reprobate minds, to ".believe a lie that they may be damned!" It is doubtless true that many of the Roman clergy possess far less intelligence than is generally awarded to them. But after spreading the broadest mantle of charity to its utmost tension, it is impossible to restrain the conviction that many of them are deliberately practicing an unprecedented fraud upon a confiding people. They certainly do know that their pretended *judicial absolution* is a blasphemous, hypocritical, ecclesiastical *farce*, and that they are willfully deceiving their unfortunate victims, and decoying them down to endless perdition. They most assuredly do know that they are instrumentally destroying the souls of their fellow-beings by crying, Peace, peace, when there is no peace for the wicked. No language can portray the consequences of this fatal deception.

The mere gratification of ambition, avarice, or lust here will be a miserable equivalent to the clergy when justice is awarded by the Judge of all the earth, who will do right; when popes, bishops, and priests, in common with other sinners, will stand justified, regenerated, and saved by grace, through faith, in the merits of Jesus Christ alone, or be forever condemned for rejecting the only Savior of the world. Lordly titles and clerical robes are not the requisite qualification for heaven. They will be consumed by the brightness of His coming. And unless clothed in the righteousness of Jesus Christ, penitent and priest, cardinal and pope, will appear destitute and naked before God, the judge of the universe, to receive merited condemnation, and ever after be exposed to that storm of wrath which is now heaping up against the day of wrath.

8. This pretended clerical power is again predicated on the assumption that Jesus Christ, in his nature and person as *man*, judicially pardoned sins; that he delegated to Peter, as *man*, power to forgive sins; and that through Peter the Roman clergy, individually, possess this power through an unbroken apostolical succession. This theory is false in every member. The Bible nowhere teaches that in forgiving sins Jesus Christ acted *only* " *as man.*"

There is no evidence in the Bible that Jesus Christ conferred on Peter, or any other apostle, judicial power to forgive sins.

The apostles never had *apostolic successors*, and the Bible does not show that Peter or any other apostle ever exercised that power, or conferred it on a successor. History does not show an unbroken *holy* succession from Peter or any other apostle to the clergy of Rome at the present time. If there is any natural or clerical affinity with any apostle and the clergy of Rome, it most legitimately connects with Judas Iscariot, whose penurious spirit they clearly manifest. There is no evidence from the Bible or history that Peter or any of the apostles went about hearing auricular confession, and forgiving sins, or that they appointed any person to do it for them. The reverse of this is true, and history and the Bible prove conclusively that the boasted "holy apostolic" succession of the Roman clergy is a myth of their own production, and fabricated for sordid purposes. History shows a long succession of popes and bishops who were clerical tyrants, and many of them drunken debauchees, by whose notorious profligacy their pretended chain of *holy* succession is hopelessly and ruinously broken.

9. This whole system rests on an unsustained *assumption*, and that assumption is enforced by the *third precept* of the Church, and the reason assigned for the *third precept* is, "Because *libertines* would not have done it [confessed] once in many years." This is a tacit admission that the Church of Rome contains within it as

communicants such a large number of libertines that it became necessary to enact a law in perpetuity to regulate their licentiousness. How different this system from the teaching of the Bible, which declares that "without *are* dogs, and sorcerers, and whoremongers!"

The Church of Jesus Christ is not the appropriate place for *libertines*, and if the Church of Rome were a true and pure Church of Christ, there would be no necessity for either the "*third precept*" of the Church, or the confessional to regulate or restrain libertines.

CHAPTER IV.

THE SEAL OF CONFESSION.

IN all ages of the world, wicked men "have loved darkness rather than light, because their deeds were evil;" and never was this fact more forcibly illustrated than in the Romish confessional. To conceal the abominations of Auricular Confession, the highest theology of the Roman Church authorizes equivocation, mental reservation, falsehood, and perjury.

For the benefit of Protestants who may not have access to the secret abominations of the confessional, we will compel Roman theologians to define the subject. The works from which we quote are now before us, and are circulated in the United States, with the approbation of popes and bishops.

Beginning with the smaller catechisms, and ending with the higher theological works, the obligation of secrecy is enforced, under the most solemn sanctions and the most awful penalties :

"The priest is bound, under pain of damnation, to absolute secrecy and silence, with regard to the sins he hears in confession." (Bishop David's Catechism, p. 105.)

"Know that the confessor is bound to suffer himself to be burnt alive sooner than disclose a single venial sin confessed by a penitent. The confessor can not speak of what he has heard in confession, even to the penitent himself; that is, without the permission of the penitent." (St. Liguori on the Commandments and Sacraments, p. 225.)

Again:

"A penitent at confession should imagine himself to be a criminal condemned to death, bound by as many chains as he has sins to confess, and presenting himself before a confessor, who holds the place of God, and who alone can loose his bonds and deliver him from hell." (p. 227.)

"By the law of God and his Church, whatever is declared in confession can never be discovered directly or indirectly to any one, upon any account whatsoever, but remains an eternal secret betwixt God and the penitent soul; of which the confessor can not, even to save his own life, make any use at all, to the penitent's discredit, disadvantage, or any other grievance whatsoever. *Vide Decretum Innocentii* XI, *die* 18 *Novemb. Anno* 1682." (Challoner's Catholic Christian Instructed, p. 126.)

"The priest, as the vicegerent of Jesus Christ, bound to eternal secrecy by every law, human and divine." (Catechism of Trent, p. 190.)

"Secrecy should be strictly observed, as well by penitent as priest; and hence, because in such circumstances secrecy must be insecure, no one can, on any account, confess by messenger or letter." (Catechism of Trent, p. 195.)

Here let us pause and sum up these facts.

Bishop David teaches that the priest is *bound* to *secrecy*, under "*pain of damnation.*" St. Liguori says that the priest should be "*burnt alive*" sooner than reveal. Dr. Challoner declares that the priest "*should not reveal, to save his life.*" The Catechism of Trent declares that the "*priest is bound to secrecy by every law, human and divine,*" and that the "*penitent is equally bound.*" Both priest and penitent are therefore bound to observe "*eternal secrecy,*" relative to transactions in the confessional, at the peril of life, and threatened with "eternal damnation" if they reveal.

Nothing less than deeds of darkness most *horrible* could demand such a penalty for disclosing their secrets.

This obligation of secrecy is not peculiar to the Roman Church in Italy and Spain, or in the dark ages of superstition. It is *now* binding on Romanists in America and throughout the world.

These pledges of eternal secrecy are not sufficient to destroy the inherent modesty enstamped by the Creator on the female constitution. The Roman clergy therefore often experience much difficulty in degrading and subjugating the noble heart of woman to the corrupting and licentious influences of the confessional. They are compelled to denounce virtuous modesty as "*foolish bashfulness*" and "*false delicacy*," and assault the citadel of the virtuous woman's heart by scoff and scorn, by threat and promise, to consummate their fiendish purpose.

In confirmation of these facts, we again refer to the Catechism of Trent, page 197:

"But as all are anxious that their sins should be buried in eternal secrecy, the faithful are to be admonished that there is no reason whatever to apprehend that what is made known in confession will ever be revealed by any priest, or that by it the penitent can, at any time, be brought into danger or difficulty of any sort. All laws, human and divine, guard the inviolability of the seal of confession, and against its sacrilegious infraction the Church denounces her heaviest chastisements. Let the priests, says the Great Council of Lateran, take especial care neither by word nor sign, nor by any other means whatever, to betray in the least degree the sacred trust confided to them by the sinner."

Again, on page 198:

"Still more pernicious is the conduct of those who, yielding to a foolish bashfulness, can not induce themselves to confess their sins. Such persons are to be encouraged by exhortation, and to be reminded that there is no reason whatever why they should yield to such false delicacy; that to no one can it appear surprising if persons fall into sin, the common malady of the human race, and natural appendage of human infirmity."

Again, on page 199:

"But as it sometimes happens that females, who may have forgotten some sin in a former confession, can not bring themselves to return to the confessor, dreading to expose themselves to the suspicion of having been guilty of something grievous, or of looking for the praise of extraordinary piety, the pastor will frequently remind the faithful, both publicly and privately, that no one is gifted with so tenacious a memory as to be able to recollect all his thoughts, words, and actions; that the faithful, therefore, should they call to mind any thing grievous which they had previously forgotten, should not be deterred from returning to the priest. These and many other matters demand, and should receive, the particular attention of the confessor in the tribunal of penance."

When we consider the nature of the questions propounded by the priests to females—maids, matrons, and small girls—it should not be a matter of surprise that the priests are compelled to tax their ingenuity in devising means by which to compel their attendance at confession. It is rather a matter of surprise that insulted virtue has so long refrained from consigning them to merited infamy.

No other class of men would be tolerated in decent society who would propound to females such *vile* questions as are asked by priests in the confessional. And yet Protestant parents, who profess to love their daughters as they love their own lives, will place them in convents, where their morals are liable to be corrupted through the unhallowed influence of the confessional. Surely, they do not know the corrupting influence to which they are exposed. But it may be said that it is a matter of discretion whether Romanists do or do not attend confession—whether they do or do not answer the obscene questions propounded by the priests.

Such declarations are evidence of the most profound ignorance of the rules of the Roman Church. It is not discretionary with any Romanist. All are *required* to make confession to the priests, and are excluded from the communion of the Church if they do not. All are required to confess their sins of thought, of word, or action—not in general terms, but in detail—and answer any questions, obscene or otherwise, which the priest may choose to ask. In attestation of these facts, we appeal to Roman books before us.

But before we proceed with the horrible disclosures, let further evidence be exhibited relative to the obligations of secrecy, by which this system of ecclesiastical seduction has been so long and so successfully secluded from inspection, and its projectors shielded from merited infamy.

Let no one infer that our language is too strong, or that we are making assertions without clear documentary evidence at hand to sustain them. We have the most horrible and startling facts before us; but their indelicacy precludes their insertion in this work. We can only approximate the facts and permit the reader to infer the rest.

The startling facts disclosed in these books and referred to in the following pages, have excited profound interest in the minds of many intelligent Protestants, and the questions are frequently asked, "*Is it possible* that such books are now secretly circulated in our midst, as a guide of the Roman clergy in the confessional and other pretended devotions?"—To which we reply, It is not only possible, but it is *absolutely certain* that they are now used on both continents with the approbation of Pope Gregory XVI., and Pope Pius IX. Peter Dens's "Theology" has been in use among the Roman clergy more than one hundred years. It has been twice unanimously approved by the Roman Catholic prelates of Ireland, during the present century, as the most complete system of theology that could be published. It has been used as a text-book in the Royal College of Maynooth, Ireland. It is secretly sold by the pope's accredited publishers and booksellers in New York. The Mechliniæ edition, from which the extracts are taken, bears date 1864, and is published by "*De Propaganda Fide*" (Society for Propagation of the Faith). On the title-page it bears the following inscription: "*Theologia ad usum Seminariorum, et Sacræ Theologiæ Alumnorum,*" (Theology in use in the Theological Seminary, and Sacred Theology for Students.) Kenrick's "Theology" was first published in Philadelphia, in the years 1841, 1842, and 1843, and "Entered according to Act of Congress, by Francis Patrick Kenrick, in the Clerk's Office of the District Court in the Eastern District of Pennsylvania." It was published in three volumes, and the extracts are from the first edition. A later edition from Mechliniæ, published in two volumes, by the "Society for the Propagation of the Faith," bears date 1861, and to our personal knowledge it is catalogued in Latin, and is for sale in the large Catholic bookstores generally throughout the United States.

These and other kindred corrupting Roman Theologies, together with Auricular Confession, ought to be suppressed by legal enactments. They are the prolific source of gross licentiousness. These are but specimens of the entire system of theology, and the infernal questions which they suggest may be propounded by bachelor priests, at discretion, to females of all ages, from "seven years" upward; and the obligation of the confession binds them under penalty of "eternal damnation" to "eternal secrecy." The indelicacy of the subjects discussed precludes the possibility of disclosing the facts promiscuously through the press or to a mixed audience. And yet something must be done to arrest this flood-tide of licentiousness.

For the benefit of those who may not have access to the original we furnish the Latin extracts on the secrecy of the confessional, accompanied with an English translation in parallel columns.

In Bishop Kenrick's Theology, vol. 3, page 172, section 87, perjury is sanctioned to conceal the abominations of the confessional, as may be seen in the following quotation:

."DE SIGILLO CONFESSIONIS.

"Interrogatus confessarius utrum quis apud eum confessus fuerit, poterit plerumque respondere, prout res se habet. Quod si clam accesserit, ipsam confessionem celatam volens, putant plures, et quidem recte, judice S. Alphonso, frangi sigillum si accessus ejus a confessario declaretur, nam gravioris, peccati suspicionem facile injicit. (L. vi. n. 638.) De iis autem quæ confitendo declarantur, nihil prorsus dicendum est; ea enim ignorare causetur; quum nonnisi Dei vices gerenti innotescant. 'Homo non adducitur in testimonium, nisi ut homo. Et ideo sine læsione conscientiæ potest jurare se nescire, quod scit tantum ut Deus.' (S. Thom. Suppl. iii. p. qu. xi. art i. ad 3.) Igitur simpliciter denegare debet se ea nosse; quod si aliunde noverit, cavendum ne quid certius ex confessione proferatur."

"THE SEAL OF CONFESSION.

"When a confessor is asked whether any one has confessed to him, he may generally reply as the case is. If he has come secretly, wishing the confession itself to be concealed, many think, and rightly, indeed, according to the opinion of S. Alphonsus (Liguori), that his seal is broken if his application to him be mentioned by the confessor, for he may easily cause him to incur suspicion of a more than commonly grievous sin. Of the things which are declared in confession, nothing further is to be said; for he is supposed not to know them when they are known only to the vicegerent of God. 'A man is brought as a witness only as a man. And, therefore, without injury to conscience, he can swear that he does not know those things, which he knows only as God.' Therefore, he ought simply to deny that he knows these things; if he has learned them from another source, care must be taken lest any thing should be reported more accurately from the confession."

Here let it be observed that the Roman priest in the confessional *is God*, and outside of the confessional, or in the court-room as a witness, *he is man.* What he, as God, knows in the confessional, he as man does not know as a witness, "*and, without injury to conscience, can swear that he does not know those things which he knows only as God.*"

Such is the moral theology of the most distinguished archbishop of America, whose works are indorsed by Pope Pius IX and Pope Gregory XVI, and are for the guide of the Roman clergy on both continents. Here is unblushing *perjury* sanctioned by the highest authority of the Church of Rome, and which all the clergy are required to teach.

The same unmitigated perjury is taught more clearly in the Moral Theology of Peter Dens. Here again is

PERJURY SANCTIONED.

"DE SIGILLO CONFESSIONIS.

"Quid est sigillum confessionis sacramentalis?

"*R.* Est obligatio seu debitum celandi ea, quæ ex sacramentali confessione cognoscuntur. (Dens. tom. vi, p. 227.)

"An potest dari casus, in quo licet frangere sigillum sacramentale?

"*R.* Non potest dari; quamvis ab eo penderet vita aut salus hominis, aut etiam interitus Reipublicæ; neque summus Pontifex in eo dispensare potest; ut proinde hoc sigilli arcanum magis liget, quam obligatio juramenti, voti, secreti

"ON THE SEAL OF CONFESSION.

"What is the seal of sacramental confession?

"*A.* It is the obligation or duty of concealing those things which are learned from sacramental confession. (Dens, vol. 6, p. 227.)

"Can a case be given in which it is lawful to break the sacramental seal?

"*A.* It can not; although the life or safety of a man depended thereon, or even the destruction of the commonwealth; nor can the Supreme Pontiff give dispensation in this; so that on that account this secret of the seal is more binding

"naturalis, etc., idque ex voluntate Dei positiva.

"Quid igitur respondere debet confessarius interrogatus super veritate, quam per solam confessionem sacramentalem novit?

"*R.* Debet respondere se nescire eam, et si opus est, idem juramento confirmare.

"*Obj.* Nullo casu licet mentiri; atqui confessarius ille mentiretur quia scit veritatem, ergo, etc.

"*R.* Neg. min., quia talis confessarius interrogatur ut homo, et respondet ut homo; jam autem non scit ut homo illam veritatem, quamvis sciat ut Deus, ait S. Th. q. II, art. 1 ad 3, et iste sensus sponte in est responsioni; nam quando extra confessionem interrogatur, vel respondet, consideratur ut homo."

"Quid si directe a confessario quæratur, utrum illud sciat per confessionem 'sacramentalem?

"*R.* Hoc casu nihil oportet respondere; ita Steyært cum Sylvio; sed interrogatio rejicienda est tanquam impia vel etiam posset absolute, non relative ad petitionem dicere;

than the obligation of an oath, a vow, a natural secret, etc., and that by the positive will of God.

"What answer, then, ought a confessor give when questioned concerning the truth which he knows from sacramental confession only?

"*A. He ought to answer that he does not know it, and, if it be necessary, to confirm the same with an oath.*

"*Obj.* It is in no case lawful to tell a lie; but that confessor would be guilty of a lie, because he knows the truth, therefore, etc.

"*A.* I deny the minor; because such a confessor is questioned as a man, and answers as a man; but now he does not know that truth as a man, though he knows it as God, says St. Thomas (q. II., art. 1, 3), and that is the free and natural meaning of the answer; for when he is asked, or when he answers outside confession, he is considered as a man.

"What if a confessor were directly asked whether he knows it through sacramental confession?

"*A.* In this case he ought to give no answer (so Steyart and Sylvius), but reject the question as impious: or he could even say absolutely, not relatively to the question, I know

ego nihil scio; quia vox *ego* nothing, because the word I
restringit ad scientiam huma- restricts to his human knowl-
nam." (Dens, tom. vi, p. 228.) edge." (Dens, v. 6, p. 228.)

Thus, in their highest theology, "perjury" is taught in the plainest possible terms. The obligation of secrecy is "*more binding than an oath, a vow, a natural secret, etc., and that by the positive will of God.*" So binding is this obligation of secrecy in the confessional that "*a case can not be given in which it is lawful to break the seal* (that is, reveal the secrets), *although the life or safety of a man depended thereon, or even the destruction of the Commonwealth.*" What horrible corruption must there be practiced in confession to require such fearful obligations of secrecy.

Reader, examine well this fiendish obligation, and understand if a Roman priest should learn in the confessional that you were to be assassinated in one hour, he dare not disclose the fact under less penalty than "*endless damnation.*" He may be your nearest neighbor; he may profess to be your personal friend; and you may have saved his life, or done him a thousand favors, but all are naught when contrasted with the more binding obligation of secrecy in the confessional. If a Roman priest should, through the confessional, learn that the Congress Hall was to be blown to atoms by gunpowder, and that the President of the United States (including the Cabinet, Congress, and visitors), were to be dashed to atoms in a moment, he dare not reveal the fact. His obligation of "eternal secrecy" binds him to silence, if "the destruction of the Commonwealth depended thereon." This fact, so emphatically set forth in moral theology, is confirmed in Roman ecclesiastical history. Priest Garnet was the confessor of the conspirators engaged in the gunpowder plot to blow up the British Parliament, with intention to destroy the royal family, that Romanists might grasp the regal power and subjugate England to Rome. He knew all the facts; he was in the confidence of the treasonable conspirators, aiding and abetting, until their fiendish plot was detected, and he lost his life for his perfidy. When arrested and convicted by a jury of his countrymen, and sentenced to be hung, drawn, and quartered, according to British law for treason, he still retained the secrets of the confessional to the last hour of life. When all hope had fled, and a certain and terrible death awaited him in a few moments, he, on the scaffold, exclaimed: "As I hope for salvation, I never was acquainted with this treasonable conspiracy except through the confessional, which I was obliged not to reveal."

This fact is found on page 580 of the "History of the Christian Church," by Rev. Joseph Reeve, with the approbation of the Right Rev. Bishop Fitzpatrick, and for sale in the large Roman bookstores generally.

St. Liguori also, and other saints of Rome, sanctioned perjury, to conceal the corrupt communications of the confessional.

In the Roman Calendar for 1845, page 167, we learn that, preparatory to his canonization, the Moral Theology of St. Liguori had been more than twenty times rigorously discussed by the Sacred Congregation of Rights, which decreed that, *in all his works*, whether printed or inedited, *not one word had been found worthy of censure;* which decree was afterward confirmed by Pope Pius VII. This Liguori was Cardinal Wiseman's favorite saint, and the following are specimens of his doctrines on the seal of confession, when the priest or penitent is interrogated relative to the secrets of the confessional:

"RESPOND 1. Sigillum hoc est obligatio juris divini strictissima in omni causa, etiam quo integri regni salus periclitaretur ad tacendum etiam post mortem pœnitentis dicta in confessione (id est in ordinead absolutionem sacramentalem), omnia, quorum revelatio sacramentum rederit onerosum vel odiosum." (Liguori, tom 6, p. 276, n. 634.)

"ANSWER 1. That this seal is an obligation of divine rght, most strict in every case, even where the safety of a whole nation would be at stake, to observe silence even after the death of the penitent as to all things spoken in confession (that is, in order to obtain sacramental absolution), the revelation of which would render the sacrament itself grievous or odious." (Liguori, vol. 6, p. 276, No. 634.)

"Quæritur an confessarius interrogatus de peccato pœnitentis possit dicere se illud nescire, etiam cum juramento. Affirmandum cum communi, quam tenent D. Thomas." (Suppl., q. 11., art. 1., ad 3.)

"It is asked whether the confessor, interrogated concerning the sin of his penitent, can say that he does not know it, even with an oath. It is answered in the affirmative, in accordance with the common opinion which St. Thomas and others hold." (Supl., q. 11. art. 1 and 3.)

The reason assigned by St. Thomas is in strict conformity to the Jesuit casuistry of Roman theologians generally, and is as follows:

"Homo non adducitur in testimonium, nisi ut homo, ideo. . . . potest jurare se nescire quod scit tantum ut Deus, (et hoc, etiamsi confessarius rogatus fuerit ad respondendum non ut homo, sed præcipue ut minister Dei, prout recte siunt Saurez et præfati auctores loc. cit.); quia confessarius nullo modo scit peccatum scientia qua possit uti ad respondendum, unde juste asserit se nescire id quod sine injustitia nequit manifestare. Vide dicta 1. 3. n. 125, v. Hinc. Quid, si insuper rogetur ad respondendum sine æquivocatione? Adhuc juramento cum potest respondere, se nescire, ut probabilus dicunt Lugo, n. 79, Croix, 1, c. cum Stoz. et Holzm. num. 722, cum Michel, contra alios. Ratio, quia tunc confessarius revera respondet secundum juramentum factum quod semper factum intelligitur modo quo fieri poterat, nempe manifestandi veritatem sine æquivocatione, sed sine æquivocatione illa, quæ licite

"A man is not adduced in testimony unless as a man; therefore, he can swear that he does not know what he knows only as God (and this holds good, although a confessor may have been asked to give his answer, not as man, but especially as minister of God, as Suarez and the before quoted authors rightly say); because a confessor, in no manner, knows a sin with a knowledge which he can use for the purpose of answering; wherefore he justly asserts that he does not know that, which, without injustice, he can not manifest. Hence, *what if he should be asked to answer without equivocation? Even in that case he can answer with an oath, that he does not know it;* as, most probably, Lugo, Croix, Stoz. et Holzm, with Michel, teach against others. The reason is, because then the confessor verily answers according to the oath made, which is always understood to be made in the manner in which it was possible to be made, to-wit, of manifesting the truth without equivocation; that is, without that equivocation which lawfully

omitti poterat: quoad æquivocationem vero necessariam, quæ non poterat omitti absque peccato, nec alter habet jus ut sine æquivocatione ei respondeatur, nec ideo confessarius tenetur sine æquivocatione respondere." (Liguori, tom. 6, n. 646.)

can be omitted. But as to the necessary equivocation, which could not be omitted without sin, the other has not a right that an answer should be given to him without equivocation; neither, moreover, is the confessor bound to answer without equivocation." (Liguori, vol. 6, n. 646.)

We have before us ten volumes (the full set) of St. Liguori's Moral Theology, from which the Roman clergy are instructed as to vile and indelicate questions in the confessional, and the manner of concealing the facts by equivocation, falsehood, and perjury; but, for the present, the above may suffice.

Again, perjury is sanctioned by De la Hogue, whose works are much esteemed, and have been in use in the Royal College of Maynooth, where Irish priests are drilled in the ritual of Auricular Confession in its filthiest details. He says:

"Si sacerdos a magistratu interrogetur de iis quorum notitiam ex sola confessione habuit, respondere debet se nescire, immo hoc ipsum jurare absque ullo mendacii periculo. Ratio est juxta Estium, quia nec mentitur, nec in equivoco ludit, qui ad mentem, interrogantis respondet, at nihil nisi verum profert; atqui ita se habet Sacerdos in prefato casu, namque ab illo non quærit

"If a priest is questioned by a magistrate as to matters which he has learned from confession alone, he ought to reply that he is ignorant of them; nay, he ought to swear to it, which he may do without *any* danger of falsehood. It is added, on the authority of Estius, that in doing so he neither lies nor equivocates, since he frames a true reply to the intention of the person interro-

Judex quid scit via confessionis *quatenus Dei vices agit*, sed quid noverit, *quatenus homo*, proindeque extra confessionem." (De la Hogue, tom. 1, p. 292.) gating him; because the magistrate does *not* ask him what he knows from confession '*in his character as God*,' but what he knows 'in his character as man,' without confession." (De la Hogue, vol. 1, p. 292.)

Adopting such theology as this, what confidence can be placed in the word or oath of a Roman priest or bishop? Under such teaching, whose character, property, or life may not be sworn away from him if the interests of papacy demand it.

The following incident may illustrate the estimate in which the laity hold the obligation of secrecy in the confessional, as reported in the *Northwestern Christian Advocate*, of 1855:

"A Roman Catholic priest was recently before a magistrate in Chicago, charged with beating and otherwise abusing a a woman, a member of his Church, for refusing to take her children from the free-school at his bidding. The defense set up was, that the transactions of the confessional were to be kept secret; that the woman knew this, and if she should violate this solemn obligation she was unworthy of belief. Witnesses, members of the Catholic Church, were examined, who testified that, according to the canons of the Church, whatever insult a priest might offer a woman at the confessional, she was bound to keep it secret from her husband."

After a thorough examination of Roman theology, we are persuaded that the Roman clergy should not be trusted under oath in any matter involving the real or imaginary interests of the Church of Rome. They claim power to absolve each other from the obligations of an oath.

This doctrine is also imparted to the laity, as may be seen by reference to St. Liguori on the Commandments and Sacraments, pages 83 and 85:

"But if, in a matter of small moment, a person swore with the intention of performing his promise, but afterward did not adhere to it, it is probable, as several theologians say, that he would not be guilty of a mortal sin; because God is called on to attest the present intention, and not the future execution of the promise."

"How is the obligation of an oath taken away? It may be taken away by annulment, by dispensation, commutation, and relaxation. First, it may be annulled by any one who has dominative power, such as a father, a husband, a guardian, prelate, or abbess; and, to annul an oath, a just cause is not necessary. Secondly, by dispensation or commutation; and such dispensation or commutation may be given by the pope or bishop, but, to grant a dispensation or commutation, a just cause is required. Thirdly, by relaxation. This may be given by the bishop, and by all who have episcopal functions."

These extracts are from a common manual in the hands of the laity. It is printed in plain English, to be read by all at discretion. The influence of such teaching, by a professedly infallible Church, on the minds of its subjects, may be easily inferred.

The most binding oaths may be violated with impunity, and with the approbation of ecclesiastical superiors.

Under the influence of these principles, it is not strange that Protestants have no rights which a Romanist is under obligation to respect, except in Protestant countries, or where papists are in the minority.

Oaths are but toys in the hand of the Roman clergy. The preceding binding obligations of secrecy are not sufficient to restrain them except at pleasure, and their theology provides for disclosures where "*a just* cause exists." The evidence is before us, but space will not permit its insertion. The solemn and repeated declaration of secrecy on the part of the clergy accomplishes, at least, three things: 1. It tends to diminish the restraints of modesty; 2. It lessens the probabilities that criminal intercourse will be exposed; 3. It furnishes incentives to yield to the seductive influences of the confessional, with the assurance that the facts will not come to light, they being known only to the guilty parties, who are supposed to be mutually interested to conceal their shame. The confiding penitent finds it difficult to believe that one so holy as the father confessor, and acting *as God* in the confessional, could so far forget his obligation of secrecy as to betray confidence, or be guilty of such perfidy, as to incur the penalty of mortal sin. And such would seem to be the fair and logical inference if things were as they seem to be, and if there were not facts, history, and theology, to the contrary. Little do the confiding common people know of the *secret* theology and Jesuit casuistry of the Roman clergy; little do they suspect that their humble and sincere confessions often furnish themes of ribaldry and jest in the carnivals and bacchanalian orgies of at least some of their lordly confessors. Dens, Liguori, and St. Thomas, each provide that, under certain contingencies, the obligation of secrecy does not bind the clergy except at discretion, and the interest of the Church must determine the matter.

The mystery of iniquity does not end here. These *men-Gods*, who, in the court-room are men, and in the confessional, Gods, may, according to their own approved theology, not only keep concubines, but, under other circumstances, flagrantly violate the law of chastity, and, at the same time, absolve their licentious accomplices.

This fact will receive attention in a subsequent chapter.

CHAPTER V.

THE CONFESSIONAL.

EACH church or chapel is usually provided with a confessional, or place for hearing confessions, which is frequently called a confession-box. The plainest form is a chair or seat, in a retired place, where the penitent may kneel beside the priest and whisper in his ear through a temporary screen. This style is not in general use in this country. With slight variations of form and structure, they are frequently about seven feet high, four feet wide, and eight feet long, divided into two equal apartments by a thin plank partition, extending across the inclosure. In this partition is usually an aperture with lattice-work, or wicket-gate, and sometimes both, about four feet from the floor, through which to whisper the most obscene communications that ever polluted the lips of mortals. Each apartment is provided with a small door, which is usually closed with shutter or curtain. This is a description of the plain chapel style of a confession-box, and intended to furnish a general idea of the leading features of all. A more aristocratic style of confession-box may be found in many of the larger churches, the exterior of which is about as large as the above described box, with the addition of another partition, forming three small boxes. The center box is for the priest, and the boxes at the right and the left for the penitents; but only one penitent should enter at a time. And to prevent the possibility of one penitent hearing the confession of another, there is a shutter or sliding board in each partition, in addition to the lattice-work or wicket, so that when a penitent enters the box on the right, the wicket on the left is closed, and the reverse, as the case may require. The middle box for the priest is so small that he, by reclining to the right or to the left, can hear the confession of a penitent in either apartment of the box. The penitent is required to kneel, with the face as near to the priest's ear at the wicket as possible, and communicate to the priest in a whisper. The necessity for putting the "*mouth*" as near the priest's ear as possible is urged from the consideration that "some penitents commit a fault by holding themselves far away from the priest, or too far to the part of the grate nearest the door of the confessional. This obliges the priest to hurt his back by stooping forward. This should not be." (Star of Bethlehem, p. 202.)

It is said to be a mortal sin for a third person to attempt to hear the secret communications of the confessional.

Another style of confession-box is, when it is built solid in the brick wall of the building, and not a ray of light can enter it except through the shutters or curtains of the doors. Confession may be made at any time or place where priest and penitent can communicate privately under the obligation of eternal secrecy. Under these circumstances, when the subjects discussed and the nature of the communications there made are understood, it will not be difficult to infer the rest. Let it suffice for the present to say, that if such communications were made by females to unmarried men under any other circumstances, they would be excluded from decent society. The evidence of this will appear in the next chapter.

EXAMINATION OF CONSCIENCE.

The manuals of the Roman Church, for the guide of the laity in confession, contain suggestive catechetical instructions, by which they are required to refresh their memory on old subjects which may be subsequently discussed in the confessional, and upon which they may be cross-examined by the priest in the confessional, as a lawyer would examine a witness in court. And the validity of the confession is made to depend upon the fidelity in examining conscience, and the unreserved disclosures subsequently made to a bachelor priest in the dark, secluded sentry-box, commonly known as the confessional. As a specimen we select a few questions for the examination of conscience on the *Sixth* Commandment in the Douay Bible, which is properly the Seventh Commandment, "Thou shalt not commit adultery." (It is called the *Sixth* Commandment in the "Garden of the Soul," and in other popish books, on account of their omission of the *second*, which forbids the worship of images or idols. They make up the number—ten—by dividing the tenth into two.) These questions are transscribed *verbatim et literatim*, with the omission of portions of two, which are calculated to suggest modes of pollution and crime that otherwise a pure-minded person would never think of. The questions are printed in *plain English*, in a popular book of devotion, issued under the direct approbation of the most celebrated Romish archbishop of America, and to be found in the hands of intelligent Romanists generally; and it is but right that Protestants, and especially those who send their daughters to Roman seminaries or convents, should know the kind of questions that will be proposed by the priests, in the secret confessional, to their wives and daughters, in case they should be induced to embrace the religion of Rome.

I must be excused for omitting the most indecent portions of the two vilest questions of the filthy list. No decent man dare pollute with them pages to be read by the people generally. The work in which they are found is but one of a class of books which may be procured at the Roman book-stores generally. The work is stereotyped, catalogued, and sold throughout the United States. The copy before us bears date 1871, and is published in "New York by *D. & J. Sadlier & Co.*, 31 Barclay street." It is the "enlarged" *American edition*, with the approbation of Dr. Hughes, in the words following:

"'THE GARDEN OF THE SOUL' HAVING BEEN DULY EXAMINED, WE HEREBY APPROVE OF ITS PUBLICATION.
"† JOHN, ARCHBISHOP OF NEW YORK,"

which is the usual official signature of that distinguished prelate.

The following are the questions, as found on pages 213 and 214:

"VI. Have you been guilty of fornication, or adultery, or incest, or any sin against nature, either with a person of the same sex, or with any other creature? How often? Or have you designed, or attempted any such sin, or sought to induce others to it? How often?

"Have you been guilty of self-pollution? or of immodest touches of yourself? How often?

"Have you touched others, or permitted yourself to be touched by others, immodestly? or given or taken wanton kisses or embraces, or any such liberties? How often?

"Have you looked at immodest objects with pleasure or danger? read immodest books or songs to yourselves or others? kept indecent pictures? willingly given ear to, and taken pleasure in hearing, loose discourse, etc.? or sought to see or hear any thing that was immodest? How often?

"Have you exposed yourself to wanton company? or played at any indecent play? or frequented masquerades, balls, comedies, etc.? with danger to your chastity? How often?

"Have you been guilty of immodest discourses, wanton stories, jests, or songs, or words of double meaning? How often? and before how many? and were the persons to whom you spoke or sung married or single? For all this you are obliged to confess, by reason of the evil thoughts these things are apt to create in the hearers.

Have you abused the marriage bed by . . . . . . or by any pollutions? or been guilty of any irregularity in order? . . . . . How often?

"Have you, without a just cause, refused the marriage debt? and what sin followed from it? How often?

"Have you debauched any person that was innocent before? Have you forced any person, or deluded any one by deceitful promises? etc., or designed or desired to do so? How often? You are obliged to make satisfaction for the injury you have done.

"Have you taught any one evil that he knew not of before? or carried any one to lewd houses? etc. How often?"

On page 216:

"IX. Have you willingly taken pleasure in unchaste thoughts or imaginations? or entertained unchaste desires? Were the objects of your desires maids or married persons, or kinsfolks, or persons consecrated to God? How often?

"Have you taken pleasure in the irregular motions of the flesh? or not endeavored to resist them? How often?

"Have you entertained, with pleasure, the thoughts of saying or doing any thing which it would be a sin to say or do? How often?

"Have you had the desire or design of committing any sin? of what sin? How often?"

Vile as these questions are, they are but as the shadow to the substance, compared with the questions in the confessional, and to the instructions in the secret Latin theology of the Roman clergy, now before us. These questions, when contrasted with the original, are white as the paper on which we write in contrast with the ink. We dare not specify the facts; and the most vivid imagination can not do justice to the subject. Let any linguist take the Moral Theology of Dens, Kenrick, Liguori, St. Thomas, and other approved theological works, which are before us, and which are *now* the guide of the clergy in the confessional and in other duties, and they will exclaim: "The half has never been told;" nor can it be, without violating every principle of decency and instinct of virtue. It is doubtless true that the promiscuous circulation of these vile theological books would corrupt any brothel on the continent. If any man of mature years doubts these facts, let him examine the original, under the general captions: "De usu Conjugii," "De Luxuria," "De Peccatis Carnalibus Conjugum inter se," "De Absolutione Complicis," "De justis causis permittendi Motus Sensualitatis," and kindred subjects which are discussed in the most minute and disgusting details, and his doubts will vanish. We have repeatedly compelled priests, in presence of large congregations of *men alone*, to admit the books and facts, and the justness of our published extracts and translations. Some of the best linguists on this continent have heard our secret lecture to *men*, and have compared the "extracts" with the original; they have declared the original books genuine, the extracts fair, the translations literal, and our strictures just.

The communications with females in the confessional are not in a dead language, nor in doubtful and obscure suggestions; but often in the most obscene vernacular tongue. Modesty is no protection. All sins *must be confessed*, and all questions propounded by the confessor must be promptly answered, otherwise the confession is a nullity, and absolution refused; thus leaving the penitent in *mortal sin*, and every moment exposed to die and be damned forever.

The *confiding* penitent is a helpless victim in the hands of an artful seducer, whose will is law, whose absolution is pardon, and whose displeasure may incur eternal perdition. No mother's eye can guard the timid, confiding daughter in the confessional. Her innocent, inexperienced, and confiding soul trembling before the august presence of one whom she is taught to believe is *God* in the confessional, and infallible in his instruction, how dare she resent an insult, or spurn his lecherous encroachments? O, that mothers could comprehend the danger of thus exposing the virture of their innocent daughters!

Then, the virtuous wife, in the absence of her husband, father, or brother, cloistered in a dark corner, under obligations of *"eternal secrecy,"* and exposed to *"endless damnation"* if she reveals, is compelled to answer questions which would seem sufficient to crimson the face of a devil, and "turn the cheek of darkness pale." O insulted virtue, hast thou no protector!

These disclosures *challenge investigation*, and if not *true* they are grossly slanderous, and we ought to be indicted for publishing them. Let the Roman clergy accept the issue if they dare, and we will compel them, on the witness-stand, to translate *worse things* from their own theology, under oath.

If priests are not corrupters of society and the despoilers of virtue, it is because they are better than their system of theology requires them to be.

To Protestant minds these startling facts may cause surprise, and some one may exclaim, Can it be possible that such things *now* exist? We emphatically answer, *Yes;* it is not only possible, but is absolutely *certain*, that this corrupt system exists in our midst, with the knowledge and approbation of the Pope and his clergy; and that papal laws and edicts stand unrepealed for the *extermination* of heretics who deny that the confessional is by divine appointment; and the Council of Trent plainly says, Let them be "*accursed.*"

To become a consistent Romanist, the first step is to *surrender* unconditionally the right of forming or expressing an opinion relative to faith and doctrine, and blindly submit to the dictation and domination of ecclesiastical superiors. Reason and common-sense must be stultified, facts and evidence ignored, before judgment and conscience will ever become the passive dupes of authority. This accounts in part for the acceptance of many absurdities taught by the Roman Church, and practiced by people who are otherwise intelligent. The fear of heresy, and its penalties, puts a quietus on many otherwise troubled minds and consciences.

Heresy is thus defined:

"*Q.* What vice is opposite to faith?
"*A.* Heresy.
"*Q.* What is heresy?
"*A.* It is an obstinate error in matters of faith.
"INSTRUC.—He is a heretic who obstinately maintains any thing contrary to the known faith and doctrine of the holy Catholic Church." (Poor Man's Catechism, p. 10.)

"A heretic is one who has an opinion; for such is the etymology of the word. What is understood by having an opinion is, following one's fancy and particular sentiment. A Catholic, without maintaining any particular sentiment, follows unhesitatingly the doctrine of the Church." (Garden of the Soul, p. 392.)

Also, "Ursuline Manual," page 504.

This blind submission is not discretionary on the part

of the victim. It is, in Roman countries, imperious and unconditional. Property, character, and life often depend upon it; and, above all, the salvation of the soul, or its utter ruin. Under these circumstances, faith is the creature of *authority*, and implicit *obedience* the perfection of piety.

In this country, the charge of heresy, or insubordination to the Roman clergy, subjects the person to excommunication from the Church, proscription, and persecution by the priest and his congregation. And this cruel persecution is enforced through the confessional, often to the great injury of business, person, and property.

Connected with this, the influence of education, often from infancy, and association, must be taken into the account. Thus hedged in by canon laws, decrees of councils, education, and association under the vigilant eye of an ecclesiastical dictator who is authorized to search out all secrets in the confessional, how abject the servitude, how helpless the victim!

Doubtless, many of the laity are to be pitied more than blamed; but who can sufficiently execrate their destroyers, who are presumed to be men of too much intelligence to be duped by their own devices? In the darker ages, they might have claimed some apology, but not now; and, especially in this Protestant country, they are inexcusable.

If any other class of men or ministers were known to have such communications with females as that practiced by the Roman clergy, they would be spurned as corrupters of society, and shunned as debauchers of the virtuous and innocent. How is it, and why is it, that such unblushing abominations have so long escaped merited rebuke? It must be from a want of information on this subject.

The facts are so astounding that men of intelligence, and often ministers of the Gospel, with the secret theology of the Roman clergy in their hands, authenticated by the most positive evidence, have exclaimed: "Is it possible! I never had the most remote conception that such things are now practiced in our midst." And the facts can not be successfully denied.

CHAPTER VI.

SINS, MORTAL AND VENIAL.

ROMAN theologians classify sins under two general divisions, usually denominated *mortal* and *venial*, by which they mean *large* and *small* sins.

Mortal (large) sins *must* be confessed; venial (small) sins may or may not be confessed. The former can only be forgiven by the clergy; the latter may be forgiven by holy water, the eucharist, penance, and various other appliances, not excepting purgatorial fire.

Original sin, which precedes both, is washed away by baptism from both infants and adults; consequently, does not legitimately come within the sphere of Auricular Confession.

In approved moral theology and catechisms, sins are defined as follows:

"What is mortal sin?
"It is that which of itself entails spiritual death upon the soul.
"What is venial sin?
"That which does not entail spiritual death upon the soul." (Dens, vol. 1, No. 153.)

This distinguished theologian devotes not less than twenty-one chapters to this important definition, marking the nice distinctions and intricacies between mortal and venial sins, and leaves the subject about as clear as when he found it.

The 156th number commences with the following words:

"Although mortal sin is far removed from venial sin, it is extremely difficult to discover, and very dangerous to define which is mortal and which venial; so that these are matters which ought to be considered, not by a human, but a divine, mind."

This *divine mind*, in his conception, evidently belongs to Roman theologians, which is shown by the fact that, immediately after this avowal of the difficulty and danger of the enterprise, he wrote twenty chapters relative to a definition of the difference. If it be thus difficult for a learned Doctor of Divinity to distinguish between *mortal* and *venial* sins, what must be the condition of the common people, who have not access to these profound theological dissertations? How shall they know what sins are mortal and what sins are not, what sins to confess and what not confess? This distinction without a difference has been a source of much perplexity to hair-splitting Roman theologians, and has caused them to darken counsel by a multitude of words.

"*Q.* What is mortal sin?

"*A.* Mortal sin is that which kills the soul, and deserves hell.

"*Q.* How does mortal sin kill the soul?

"*A.* Mortal sin kills the soul by destroying the life of the soul, which is the grace of God.

"*Q.* What is venial sin?

"*A.* Venial sin is that which does not kill the soul, yet displeases God." (General Catechism, p. 18.)

"*Q.* What if one willfully conceal a mortal sin in confession?

"*A.* He who conceals a mortal sin in confession commits a great sin by telling a lie to the Holy Ghost, and makes his confession worthless.

"*Q.* What must we do that we may not be guilty of leaving out sins in confession?

"*A.* That we may not be guilty of leaving out sins in confession, we must carefully examine our consciences upon the Ten Commandments, the seven deadly sins, etc." (p. 41.)

"*Q.* How many are the chief mortal sins, commonly called capital and deadly sins?

"*A.* Seven: Pride, Covetousness, Lust, Anger, Gluttony, Envy, Sloth.

"*Q.* Where shall they go who die in mortal sin?

"*A.* To hell, for all eternity.

"*Q.* Where do they go who die in venial sin?

"*A.* To purgatory." (Butler's Catechism, p. 27.)

These are specimens of mortal sins which may be enlarged indefinitely. It is a mortal sin to read the Bible, to attend a Protestant Church, or to read books published by Protestants, or to form opinions contrary to the known faith and doctrine of the Roman Church. It is a mortal sin to join the Odd Fellows or Masons, or any forbidden society. It is a mortal sin not to *confess* and *pay* the priest.

Falsehood, perjury, theft, arson, and murder, *may* or *may not* be charged as *mortal* sins. Circumstances must determine these matters. It is as true now in that system as it was in the days of the Inquisition, that the end sanctifies the means. The *greater good* to the Church of Rome must decide these vexed questions.

This distinction between *mortal* and *venial* sins furnishes a wide field for Romanist casuistry, and leaves the confiding penitent in doubt as to his present condition and his final destiny. *Mortal* sins only *must* be confessed. The Council of Trent says, Chapter V, 16:

"For venial offenses, by which we are not excluded from the grace of God, and into which we so frequently fall, may be concealed without fault, and expiated in many other ways."

Mortal sins, even of thought, make men children of wrath, and enemies of God, and must be exposed in minute detail, with all the attendant circumstances which may aggravate or palliate the offense. And for this the Council of Trent assigns the following reason:

"It is plain that the priests can not sustain the office of judge, if the cause be unknown to them; or inflict equitable punishments, if sins are only confessed in general, and not minutely and individually described. For this reason it follows that penitents are bound to rehearse in confession all mortal sins, of which, after diligent examination of themselves, they are conscious, even though they be of the most secret kind, and only committed against the two last precepts of the Decalogue, etc. . . . Moreover, it follows, that even those circumstances which alter the species of sin are to be explained in confession, since otherwise the penitents can not fully confess their sins, nor the judge know them." (Ch. v.)

"Though the priest's absolution is the dispensation of a benefit which belongs to another, yet it is not to be considered as merely a ministry, whether to publish the Gospel or to declare the remission of sins, *but as of the nature of a judicial act, in which sentence is pronounced by him as a* JUDGE." (Ch. 6, of the Minister.)

The priest who hears confession is represented as sitting in the tribunal of penance as Christ himself, as a judge forgiving sins and inflicting punishment. This is the orthodox faith of the Church, and when denied, is the result of ignorance, or a matter of expediency to conceal the facts. The Catechism of Trent says:

"The absolution of the priest, which is expressed in words, seals, therefore, the remission of sins which it accomplishes in the soul." (P. 180.)

"Unlike the authority given to the priests of the Old Law, to declare the leper cleansed from his leprosy, the power with which the priests of the New Law are invested is not simply to declare that sins are forgiven, but, as the ministers of God, *really to absolve from sin;* a power which God himself, the author and source of grace and justification, exercises through their ministry." (P. 182.)

"There is no sin, however enormous, or however frequently repeated, which penance does not remit." (P. 183.)

"The voice of the priest, who is legitimately constituted a minister for the remission of sins, *is to be heard as that of Christ himself,* who said to the lame man 'Son, be of good cheer, thy sins are forgiven thee.'" (P. 189.)

This may all seem clear to those who *may not* investigate, and who dare not doubt; but what are the facts, and where is the evidence to sustain them? In this system things are assumed which most need proof to sustain them.

Where is the evidence that priests, either good or bad, can infallibly discriminate between mortal and venial sins, and, as God, grant judicial pardon, or retain sin? And what presumptuous mortal dare assert that to violate the moral law and offend an infinite God is only a *venial sin?* Who shall decide this momentous question, involving the destiny of immortal souls?

Where is the authority from the Bible for this absurd division of sins into *mortal* and *venial*—the former deserving endless punishment, and the latter temporal punishment; the former only forgiven by a Roman priest, the latter by other devices, including holy water, the eucharist, penance, and purgatorial fire? We are informed that there are just *"seven chief mortal sins:"* "Pride," "covetousness," "lust," "anger," "gluttony," "envy," "sloth." Now, this is true, or it is not true. If true, the evidence of its truthfulness will be found in the Bible. So important a matter will not be left to inference or conjecture. It is a personal matter with every intelligent being on earth. *One* mistake here, according to this doctrine, may destroy the soul forever.

But it is a singular fact that inspired men, engaged in writing the Scriptures during a period of more than fifteen hundred years, and discussing sin in all its forms and phases, never made this modern discovery relative to the great distinction between *mortal* and *venial* sins. They never marked the line of distinction where *venial* sin becomes *mortal*, or the *finite* becomes *infinite*. This important omission has been a great source of annoyance to Roman theologians, and will probably so continue till their system of theology changes for the better, or till it is numbered with the things of the past. How the Roman clergy ascertained that there were *seven* mortal sins, we are not informed; we have this assertion, and nothing more. But at every step we encounter difficulty. Pride, for example, is a deadly sin. Is *every degree* of pride deadly or mortal sin? If so, all who are not perfect in humility are constantly living in mortal sin. If this be true, how many of the clergy would be free from mortal sin for one hour? Would the Pope of Rome escape? But if every degree of pride is not mortal sin, in what degree must it exist before it becomes deadly? When does it pass that undefined, intangible line which separates the venial from the mortal sins? Echo answers, where? Here we are left in the dark. All is indefinite.

Again, we are informed that *covetousness* is a deadly sin. Is every degree of it so? If not, what degree is? The same questions may be asked relative to the whole seven mortal sins, and no definite answer given. The division of sin into mortal and venial is absurd. After all, if the list of mortal sins be admitted, we are not sure that it is complete. *Lying* and *stealing* are usually regarded as sins. Is every lie a mortal sin? If so, what would become of the clergy? And if not, how many lies constitute a *mortal* sin? The Bible says: "Thou shalt not steal." But some persons do steal, which is a positive violation of the law of God. Is every theft a mortal sin? Or is the violation of a commandment of the Decalogue only a small sin which may be washed away by a few drops of holy-water, or a few alms-deeds, or a little penance? Here St. Liguori answers this question conditionally. He says of the thief:

"If he has taken a valuable material at any one time, he has sinned mortally at that time. But if he has stolen a small amount at different times, then he has not sinned mortally, unless it amount to a valuable quantity; provided that from the beginning he had not the intention of reaching a valuable amount; but since that amount has now become considerable (*gravis*), although he has not sinned mortally, yet he is bound—*sub-gravi*—under mortal sin, to restitution, *at least of that last quantity which constituted the amount considerable.*"

According to this distinguished theologian and saint, a man may steal small quantities without being guilty of mortal sin; and when goods or money stolen amount to a considerable sum (the language is indefinite), he is bound to restore *the last quantity stolen.* Of course he may keep the rest, and only be guilty of a venial sin, which is a small matter, and may, or may not, be mentioned in the confessional. Again he says:

"But probably those who have eaten fruit in the vineyards of others, provided they be not rare, or of great price, may be excused, at least from mortal sin, if they do not carry it away in large quantities. For in things of this kind, which are too little expounded, a greater quantity is required to constitute a valuable amount. And in this way men-servants and maid-servants may be easily excused, who take from their master's tables, provided they be not in large quantities, or extraordinary. Neither ought those to be regarded as guilty of mortal sin who cut wood, or take their flocks to feed in the fields of the community, though it be prohibited, because such prohibitions are supposed to be penal."

Stealing is, therefore, admissible, provided it is not in large quantities at one time. It is a small matter for servants to steal, provided they do not take too much at one time (which might lead to their detection). This possibly may, in part, account for the incessant small stealing for which many of their servants are so notorious. Again:

"When thefts are committed by children, or by wives, a much greater quantity is required to constitute the sin mortal;

and rarely are these held under strong obligation (*gravi obligatione*) to restore."

Comment is useless; theft is sanctioned, and the amount indefinite. It is *a much greater quantity* than some other quantity, but no definite specification in either case.

Again:

"If he [the thief] can not make restitution without reducing himself to severe want; that is, without falling from that state which he has justly acquired, then he may defer restitution, provided the loser be not in severe want. Nay, though the loser be in severe want, probably even then the debtor is not bound to restitution, when he is likewise in severe want, and by restitution would be placed, as it were, in extreme necessity. This, however, is understood, provided the thing stolen does not exist in species, and provided the loser was not reduced particularly by the theft to that severe necessity."

Circumstances must determine whether the property stolen must be restored.

"If the theft is uncertain, that is, if the person injured is uncertain, the penitent is bound to restore, either by causing masses to be said, or giving alms to the poor, or giving it to pious places; and if he is poor, he may apply it to himself or his family. But if the person is certain, restitution should be made to him: wherefore it is indeed wonderful that there are found so many confessors so unskillful, who, when it is known who the loser is, impose on their penitents, that for the thing to be restored they should give alms, or cause masses to be celebrated."

Here is a genuine Roman Catholic process of restitution. A theft has been perpetrated by a Romanist, and the fact known to the priest—is the thief denounced, or excluded from the Church? *No.* If he is not certain as to the person injured, he *is bound to restore, either by causing masses to be said, or giving alms to the poor, or giv-*

ing it to pious places. That is, in plain English, give it to the priests. This is not only authorizing theft, but requiring the thief to divide with the priest. And facts may be exhibited to show that this villainous practice is now sanctioned, and furnishes a source of revenue to the Roman Church. Among the thousands of Roman Catholics annually convicted of theft, who throng the house of correction, the county jail, and State-prison, not one in a thousand was ever known to make restitution to a Protestant. The question arises, What becomes of the property? Does the thief appropriate it to his own use, and conceal the fact from the priest in confession? If so, how can the priest grant valid absolution? Or does he confess, and pay it to the priest for masses? Or do they divide the stolen property between them? These are nice points in Romish theology, and Protestants demand answers.

What right has a priest, more than any other man, to conceal stolen property, or to appropriate it to his own use, or that of his Church? This subject requires investigation, and justice demands that, where a confessing Catholic is convicted of theft, and does not make restitution, the priest should be held for his knowledge of the crime and complicity in the act. The priest has knowledge of the crime, or he has not. If he has *not*, all his pretended *absolutions* are impositions, and he is *obtaining money under false pretense.* If he has knowledge of the crime, he knows where the stolen property is, and to whom it belongs, and if he does not restore it, or cause it to be restored, he is, by implication, "*particeps criminis.*" He has, in equity, no more right to screen himself from punishment under the sanctimonious garb of his profession, than any other felon. Thousands of families who reside in cities can testify that thefts among Romanists are of daily occurrence. The police court of any city will attest this fact. From the small articles in the wardrobe, bed-chamber, and cupboard, to gold watches, bracelets, and pocket-books—nothing is safe. And so far as our observation has extended (with few exceptions), the more zealous the penitent in attending confession the more frequent the thefts. These facts may not be successfully denied. It will be observed in the last extract from St. Liguori (that dear saint of blessed memory) that there is one contingency which may deprive the priest of the stolen property, that is, if the thief " is poor he may apply it to himself or his family." And history establishes the fact there is no scarcity of *poor* thieves where the Romish clergy exercise a controlling influence. Look at Italy, Spain, Mexico, and other Romish countries, where the streets are thronged with Roman Catholic paupers. And in our own country nearly all the itinerant beggars are of Romish origin. With these facts before us, will any sane man pretend that Romish schools are adapted to the wants of American youth? It requires not the wisdom of Solomon to predict that children trained under such principles are liable to be corrupted—ruined.

The worst of liars began their downward course by

telling lies which they considered of trifling importance. The worst thieves and robbers began their course by stealing small quantities. If children are taught to regard such lying and thieving "a small and very pardonable offense," they may be induced to yield to temptation, which will terminate in disgrace and ruin.

Such morality Romanists must teach in their schools and in the confessional or discard their own doctrines, which they profess to believe are infallibly true. But such is not the training required by American youth and citizens.

This division of sins into *mortal* and *venial* is grossly absurd, and more grossly immoral, and doubtless accounts, in part, for the prevalence of immorality in Roman countries, and among Romanists in Protestant countries. Dishonesty is the legitimate result of such teaching.

God says, "Thou shalt not steal." The priest says you may steal and give it to him for saying masses. Thus, by precept and example, making void the law of God. The Bible teaches that all unrighteousness is sin, and "the wages of sin is death." It does not say the wages of *mortal* sin is death. Ezekiel declares "the soul that sinneth it shall die." He did not say the soul that sinneth *mortally* shall die. If a man is a thief, he is so at heart; and whether he steal one dollar or ten thousand, he is morally a thief, and would be so at heart if there was not a dollar in the universe to steal. The impure stream only proclaims the quality of the fountain. Sin is estimated not by weights and measures, or by dollars and cents, but by the nature of the law violated and the majesty of the Being offended. All sin is against God and his law; and if ever pardoned, God must do it. Away with the blasphemous jugglery of self-constituted judicial dictators and clerical pretenders.

CHAPTER VII.

POWER OF THE KEYS.

TO make a show of decency, and to justify the abominations of the confessional, the Roman clergy, with accustomed audacity, have quoted and perverted the Scriptures. In proof of this system, we are referred to Numbers v, 6, 7, Matthew iii, 6, Acts xix, 18, and James v, 16, as authority for the confessional. But what relation have these Scriptures to the subject? Do they teach the duty of Auricular Confession? Do they define the seal of confession, which enjoins eternal secrecy? Do they affix the penalty of "eternal damnation" to be inflicted on either priest or penitent who shall reveal the secrets of the confessional? Do they prove that "*all are obliged* to confess to the priests at least once each year?" Do they show that the priest, in the confessional, "is as God," and in the court-room "is as man?" Do they prove that Auricular Confession constitutes any part of Christian duty, or that, under any circumstances; a penitent may " behold in the priest the person of Jesus Christ?" Do these Scriptures teach that Christian men and women should crouch in self-abasement before a Romish priest—regarding themselves as criminals before their judge?" Do they say any thing about "*mortal*" and *venial sins*," or about *absolution* from sin "by the power of the keys?" No; not one word about all this; not so much as a form of prayer to the Virgin Mary, nor a word of instruction about the use of "*prayer-beads.*" Truly, Moses, John the Baptist, Luke, and James were poor specimens for Roman priests. They seemed to be entirely ignorant of Auricular Confession, and nowhere inculcated the practice, either by precept or example.

But, since Romanists refer to these Scriptures as authority for their abominations in the confessional, each shall receive a separate examination.

"Speak unto the children of Israel, When a man or woman shall commit any sin that men commit, to do a trespass against the Lord, and that person be guilty: then they shall confess their sin which they have done," etc. (Num. v, 6, 7.)

Now, let it be observed that this instruction was given to *Israel*, and related to ceremonial restitution; and neither furnishes precept nor example for Auricular Confession.

The context shows that reference is here made to certain fraudulent transactions for which restitution was due; and the confession was intended to show why, and for what, the indemnity was offered. This text does not speak of confessing in the *ear* of a priest in secret. A man or woman was not required to ransack every corner of the conscience, as papists are, and in detail enumerate every evil thought, word, or deed. The confession may have been made to God, or it may have been made to the party wronged. But if it be admitted that it was made to the Jewish priest, there is no evidence that it was *whispered in his ear* in a dark corner of the Tabernacle, or any pledge of secrecy imposed on either party. The common people did not enter into the Tabernacle; but met the priest in the outer court, where it was impossible to communicate privately.

Again, with an air of confidence, we are referred to Matthew iii, 6 :

"And were baptized of him [John] in Jordan, confessing their sins."

This certainly is a bad specimen of Auricular Confession.

1. John the Baptist was a *Jewish*, and not a Romish, priest. He was also the *first-born* son of a high-priest, and consequently a high-priest himself. His father was a *married* man, and certainly not good authority with Romish priests.

2. The people confessed not in *secret* to John the Baptist, but in the presence of the multitude "from Jerusalem, and all Judea, and all the region round about Jordan." (Matthew iii, 5.)

This confession was not whispered in the ear, under a mutual pledge of secrecy; but open, free, and voluntary. It was not in a dark corner, nor in the confession-box; but at the River Jordan, and in presence of all the people. There is no record to prove that the people fell down upon their knees before "father" John, made the sign of the cross, "kissed the good priest's hand," recited the "Confiteor," and *whispered* their confessions in his ears. Not one word is said about "penance" to be performed, after confession or absolution "by the power of the keys." They came publicly, and confessed their sins; and John publicly required them, as an evidence of sincerity, to "bring forth fruits meet for repentance." They desired to "flee from the wrath to come;" and John directed them to *believe on Christ*. (Acts xix, 4.)

Again, we are referred to Acts xix, 18:

"And many that believed came, and confessed, and showed their deeds."

This text, also, fails to prove Auricular Confession.

1. It may be observed that *many*, not all, came and confessed.

2. They *showed their deeds*. Their confession became a matter of public record; it was known not only to Paul, but Luke published it to the world; and yet there is no evidence that he was excommunicated, or consigned to endless perdition.

"Many of them also which used curious arts brought their books together, and burned them before all men: and they counted the price of them, and found it fifty thousand pieces of silver."

They did not burn their books in the confession-box; for it is distinctly stated that they *burned them before all men*. Books valued at fifty thousand pieces of silver would produce an unpleasant amount of smoke in a modern confession-box. And yet there is as much evidence that they burned their books in secret as that they confessed their sins to Paul in secret. The truth is,

Paul preached to the people, not in an unknown tongue, but in a language which the common people could understand, "and the name of the Lord Jesus was magnified." Men were converted by the power of God, through the truth. They came and *showed their deeds*—exhibited their past folly as beacons to warn others. They renounced their former practices, and publicly espoused the cause of Christ.

Again, we are referred to James v, 16:

"Confess your faults one to another," etc.

This text, also, proves too much for those who advocate Auricular Confession. It does not say, Confess your faults to the *priests*, and regard yourselves as *criminals;* but it does say, Confess your faults "*one to another.*" Now, if this refers to Auricular Confession, it requires the priest to confess to the penitent, and the penitent to confess to the priest. The obligation is as binding on the one as on the other. And, after all, there might arise a question as to who should grant absolution. We have learned, from Peter Dens's "Moral Theology," that the priest in the confessional may break the Seventh Commandment (sixth of the Douay Bible), and immediately grant his accomplice in guilt absolution from "all other sins," "theft and homicide" not excepted. And in case that his accomplice in guilt is in danger of death, the priest can also grant her absolution from the sin of fornication or adultery with *himself*. But where there is not present danger of death, the case of his accomplice must be referred to another priest for absolution. It does not require extraordinary sagacity to see how two licentious priests could confess their faults *one to the other*, and each pronounce on the other absolution. This would not be more difficult than to absolve an accomplice; and, under the circumstances, might entirely dispense with penance. If the priest guilty of adultery or fornication may grant absolution to *his accomplice* in the crime, we see no good and sufficient reason why he may not grant absolution to himself also. This certainly would supersede the necessity of confession on his part, since he already knows the facts and circumstances of the case.

But, after all, it appears, from Dens's "Theology," that there are certain restrictions upon the immorality of the priests. They are required to exercise their gifts in moderation; and the priest who "deliberately falls" only "two or three times a month" ought to *doubt* his qualification for the holy office of confessor. Thus it may be seen that, in *theory* at least, all sense of decency and propriety is not wholly excluded from sacramental confession.

Having failed to find Scripture proof for Auricular Confession, the priests of Rome have recourse to inferential evidence predicated upon *false* assumptions:

1. They *assume* that the Apostle Peter was the vicar of Jesus Christ on earth; that he held the keys of the kingdom of glory, with power and authority to open heaven or hell at discretion.

2. They *assume* that they are the apostolic successors

of Peter; and, by virtue of their relation to him, they possess judicial power to forgive sin, or to retain sin; and that the kingdom of glory can only be entered by their permission.

3. They *assume* that, in order to obtain the remission of sin, all *mortal sins* must be confessed to them in secret.

4. They *assume* that all, of every creed and nation, who will not bow the knee to them in confession are to experience endless perdition.

These arrogant assumptions are set up by men who, according to their own admission, are liable to break the Seventh Commandment *two or three times a month;* and whose history furnishes many melancholy proofs of their wickedness in this respect.

Again, they refer to Matthew xviii, 19:

'Verily I say unto you, Whatsoever ye shall bind on earth shall be bound in heaven; and whatsoever ye shall loose on earth shall be loosed in heaven."

This Scripture also fails to furnish inferential proof of their assumptions. The context shows that it referred to the *discipline* of the Church by the apostles on earth; and, when administered by them according to the principles of their religion, it would meet the approbation of Heaven. The apostles neither claimed nor exercised the power of which Romish priests boast. If the power which they claim had been given to Peter, there is no evidence that he was authorized to transmit it to any other person. But there is no evidence that either

Peter or any other apostle received or exercised the power to forgive sins.

Our Lord addressed not Peter alone, but all the apostles: "Whatsoever ye shall bind," etc. He did not say *whosoever*, but *whatsoever*. He referred to things, and not to persons; to the *discipline* of the Church, and not to the *destiny* of its members.

The language is plain, "*Ye shall bind.*" The phrase *to bind* and *to loose*, among the Jews, often signified nothing more than *to prohibit* and *to permit*. To *bind* a thing was to forbid it; to *loose* a thing was to allow it to be done, and on that occasion the phrase was without doubt employed in this sense. Thus, relative to gathering wood on the Sabbath-day, they said, "The School of *Shammei binds it*"—that is, *forbids it;* "the School of Hillel, *looses it*"—that is, *allows it*. The phrase "kingdom of heaven" is frequently employed to denote the Church of Christ on earth. Matt. iii, 2, "The kingdom of heaven is at hand."

John the Baptist did not mean the kingdom of glory. No, he referred to the new dispensation into which the Church was about to enter. So in the parables of our Lord, "the kingdom of heaven is like unto a grain of mustard seed;" "the kingdom of heaven is likened unto a man which sowed good seed in his field;" it "is like leaven which a woman took and hid in three measures of meal;" it "is like unto a net." Again, the disciples asked "who is greatest in the kingdom of heaven?" (Matthew xviii, 1.) "From the days of John the

Baptist until now, the kingdom of heaven suffereth violence, and the violent take it by force." (Matthew xi, 12.)

These and other Scriptures evidently refer to the Church under the new dispensation. The apostles of Christ were ministers of *the Church on earth*. They were not priests to offer sacrifice. They were not dictators to lord it over God's heritage. They were not judicial "vicars" of Jesus Christ to consign men to perdition at pleasure. They were not as Gods in confession-boxes to forgive sins. They were ministers of Jesus Christ on earth, authorized to preach the Gospel, to admit to Church privileges those who ought to enter, and exclude the unworthy. They were authorized to admit those who gave evidence of piety, and exclude others, and the legitimate exercise of this power would meet the approbation of Heaven.

Again we are referred to John xx, 22 and 23:

"And when he had said this, he breathed on *them*, and saith unto *them*, Receive ye the Holy Ghost; whosesoever sins ye remit, they are remitted unto them; *and* whosesoever *sins* ye retain, they are retained."

This is a parallel Scripture with Matthew xviii, etc., and inculcates the same doctrine. Did he breathe on Peter alone? No, "he breathed on *them*, and saith unto *them* [to those who were present, Judas and Thomas only were absent, but probably were both present on the former occasion], Receive ye the Holy Ghost." This was a pledge of the miraculous endowment experienced by them on the day of Pentecost. (Acts ii, 1 and 2.) "Whosoever sins ye remit," etc.

The meaning of this Scripture is not that men can forgive sins, but that the inspired apostles, in founding the Church under the new dispensation should be taught by the Holy Ghost to declare *on what terms, to what characters, and to what moral state of mind* God would bestow the forgiveness of sins. They were by *inspiration* authorized to establish in all the Churches the condition on which men might be pardoned, with the assurance that all who would comply should have the evidence of forgiveness and reconciliation, and those who would not comply with the condition, should not be forgiven, but be rejected on account of their willful rejection of offered pardon.

Again, we are referred to Matt. xvi, 18:

"And I say also unto thee, that thou art Peter, and upon this rock I will build my Church; and the gates of hell shall not prevail against it."

This is the chief corner-stone in the temple of popery, and without it the building will fall to ruin. This Scripture is pressed into almost every sermon, in advocacy or defense of popery. It is on the lips of papists everywhere, and often quoted by those who never read it, and who could not read it if they had access to a Bible depository containing all the languages of earth. As to the fact that it is the language of Jesus Christ, none will deny; but the great question is, What does it mean? What was the great lesson intended to be taught? Is it true that Jesus Christ was about to establish popery, and took this occasion to announce the appointment of Peter the *first pope*, with power to continue his succession to the end of time? Did he then and there intend to appoint him Vicar-general, with "*divine right*" to govern the world, to dictate to kings, emperors, governments, and states? to preside over the Church; to hold the keys of heaven and hell, and, in person or by proxy, save or damn men at pleasure? If this was the place of Peter's coronation it must have been attended with far less pomp and parade than that of many of his pretended successors. And in imparting such extraordinary power to be perpetuated by successors to the end of time, the commission will be found clear and unequivocal, and the instructions detailed and specific. What are the facts?

When driven from every other subterfuge, Romanists appeal, in vain, to the Bible to sustain their system. They use the Scripture as Satan did on the mountain and on the pinnacle of the temple. They quote it that they may pervert it. If they really believe that the Scriptures are a sufficient rule of faith and practice, why are they continually appealing to tradition and the Church as authority? Or if we must *obey the Church*, regardless of our convictions of truth, as revealed in the Scriptures, why refer to the Scriptures? Why not go to the *Church* at once? Why compel us to form *opinions*, and thereby become *heretics*, in order to believe Romish Infallibility and Auricular Confession?

But since they have appealed to the Scriptures they shall go to their favorite text:

"Thou art Peter, and upon this rock I will build my Church, and the gates of hell shall not prevail against it. And I will give unto thee the keys of the kingdom of heaven; and whatsoever thou shalt bind on earth shall be bound in heaven: and whatsoever thou shalt loose on earth shall be loosed in heaven." (Matt. xvi, 18, 19.)

The disciples were interrogated relative to their faith in Christ: "Whom say ye that I am? Peter answered, Thou art the Christ, the Son of the living God. And Jesus answered and said unto him: Blessed art thou, Simon Bar-jona: for flesh and blood hath not revealed it unto thee, but my Father which is in heaven. And I say also unto thee, thou art Peter," etc. This Scripture is evidently the main pillar by which they attempt to prop their assumed infallibility. They apply this Scripture as if Christ had said, "Thou art Peter, and upon *thee* will I build my Church." Peter, and not Christ, is thus constituted the foundation of the Romish Church." Upon this exegesis the infallibility of the Church is boldly asserted. This is the rock on which that sect is built. Now let its solidity be tested by a few blows from the hammer of truth, and this sandstone of would-be infallibility will crumble to dust. Truth is consistent with itself. One Bible truth never contradicts another. The rules of interpretation require that any given passage in any writing is to be understood in harmony with the whole. No single paragraph or passage is to be so construed as to clash with, or contradict the uniform sense of the author on the same subject. This will apply in the case before us. Christ said to Peter, Upon *this* rock I will build my Church. The question is, To what rock did Christ refer? Was it to Peter, a short-sighted erring, fallible man? Or was it to Christ, the object of Peter's faith? Some suppose *this rock* referred to Peter; others to Peter's profession; others refer it to Christ himself; while other learned linguists regard the declaration of Christ as plain and unequivocal, and that it should be read, Thou art *Petros* (masculine gender), a rock, a stone, a pebble (movable). On *this Petra* (feminine gender), a rock, a granite, immovable, will I build my Church, and the gates of hell (councils of wicked men and devils) shall not prevail against it—shall not overcome, conquer, or subdue it.

It is admitted that *Petra* is a Greek noun in the *feminine* gender; the pronoun "*taute,*" (*this*) in the Greek text, is in the feminine gender agreeing with the noun "*Petra.*" And Petros (Peter) is in the masculine gender. *Petra*, then, must refer to something different from Peter. If the Savior had proposed to build his Church on Peter, he would have used Petros twice instead of Petros and Petra.

Can it be possible that the Omniscient Jesus, who knew the end from the beginning, and who knew the hearts of all men, would use such language on a subject so important that not one of his disciples understood it? Or that he gave to Peter the keys of heaven and hell, and that Peter was so stupid that during his life he went about with the keys dangling to his girdle, and neither he nor any other person ever suspected that he had them. If Peter had the keys, and was so ignorant of his power, or otherwise so derelict in duty that he never used them, he must have been a poor specimen of pebble on which to build a Church. In a commission in perpetuity to save or damn men, to open and shut the kingdom of glory at pleasure, obscurity or ambiguity is not admissible. A just and holy God would not thus trifle with his creatures. And it is not less unreasonable to suppose that the Omniscient Jehovah would send his Son to die for the world, that all might be saved, and then impart to a fallible man, or men, the keys of glory by which they may at pleasure exclude those for whom Christ died.

CHAPTER VIII.

THE CLERGY AND CONCUBINES.

"DE ABSOLUTIONE COMPLICIS.

"ADVERTENDUM quod nullus confessarius, extra mortis periculum, licet alias habeat potestatem absolvendi a reservatis absolvere possit aut valeat a peccato quolibet mortali externo contra castitatem, complicem in eodem secum peccato.

"Hic casus complicis non collocatur, inter casus reservatos, quia episcopus non reservat sibi absolutionem, sed quilibet alius confessarius potest ab eo absolvere, præterquam sacerdos complex." (Dens, tom. 6, 297.)

"An comprehenditur masculus complex in peccato venereo v. g. per tactus?

"R. Affirmative, quia Pontifex extendit ad qualemcumque personam.

"Non requiritur ut hoc peccatum complicis patratum sit in confessione, vel occasione confessionis; quocumque enim

"ON THE ABSOLUTION OF AN ACCOMPLICE.

"Let it be observed that, except in case of danger of death, no confessor, though he may otherwise have the power of absolving from reserved cases, may or can absolve his accomplice in any external mortal sin against chastity committed by the accomplice with the confessor himself.

"*This case of an accomplice is* NOT *placed amongst the reserved cases, because the bishop does* NOT *reserve the absolution to himself; but any other confessor can absolve from it, except the priest who is himself the partner in the act.*" (Dens, vol. 6, p. 297.)

"Is a male accomplice in venereal sin—to wit, by touches—comprehended in this degree?

"*A*. Yes; because the Pope extends it to whatsoever person.

"It is not required that this sin of an accomplice be committed in confession, or by occasion of confession; for, in

loco vel tempore factum est, etiam antequam esset confessarius, facit casum complicis." (Dens, tom. 6, 298.)

"Nota ultimo, cum restrictio fiat ad peccata carnis, poterit confessarius complicem in aliis peccatis, v. g. in furto, homicidio, etc., valide absolvere." (Dens, tom. 6, 298.)

"Confessarius sollicitavit pœnitentem ad turpia, non in confessione, nec occasione confessionis, sed ex alia occasione extraordinaria: an est denuntiandus?

"*R.* Negative. Aliud foret, si ex scientia confessionis sollicitaret; quia, v. g., ex confessione novit, illam personam deditam tali peccato venereo." (P. Antoine, tom. 4, 430; Dens, tom. 6, 298.)

whatever place or time it has been done, even before he was her confessor, it makes a case of an accomplice." (Dens, vol. 6, p. 298.)

"Lastly, take note that, since the restriction is made to carnal sins, the confessor will be able to give valid absolution to his accomplice in other sins; namely, in theft, in homicide, etc." (Dens, vol. 6, p. 298.)

"A confessor has seduced his penitent to the commission of carnal sin, not in confession, nor by occasion of confession, but from some other extraordinary occasion: is he to be denounced?

"*A.* No. If he had tampered with her from his knowledge of confession, it would be a different thing; because, for instance, he knows that person, from her confession, to be given to such carnal sins." (P. Antoine, vol. 4, p. 430; Dens, vol. 6, p. 298.)

The above extracts are from the "Mechliniæ" edition of Dens's "Moral Theology," bearing date A. D. 1864, in seven volumes. It is published by the Society for the Propagation of the Faith, and sold by the Pope's publisher and importer in New York. The original books are before us. We also have the Dublin edition, eight volumes; and the extracts here inserted will be found in it, in volume 6, pages 218 and 219. We thus minutely describe, that the clergy may not quibble as to the authenticity of the works, as they are disposed to do when there is the possibility of a doubt. If prudence would permit, we might add largely to these quotations from this and other theological works.

Here let the reader pause and consider the import of these brief extracts, and infer, if he can, what would be the condition of society and the destiny of our beloved country if this intolerant system of corruption should gain the ascendancy. A glance at the subject will disclose the following facts:

1. That a priest in the confessional, "*as God*," pretending to forgive sin, may seduce his female penitent, and pardon every other sin which she has committed except that one act of licentiousness with her lordly confessor; and, if she is in danger of death, he can pardon that sin also.

2. That when the priest, in confession, has seduced a female penitent who is not in present danger of death, he can forgive every other sin which she has committed except that act of licentiousness with himself, and another priest must pardon (or absolve) that.

This, certainly, is a very easy way to dispose of the vilest crimes. Priests may thus aid each other in their pious work of prostitution; and such female penitents may make short work of confession, seeing the facts are already known to the father-confessor who is her accomplice in guilt.

What more God-dishonoring, heaven-daring insolence

could have been devised? And yet, in the estimation of this learned Doctor of Divinity, it is not a matter of sufficient importance to be referred to the bishop. "*Any other priest can grant absolution.*"

Such is Roman theology and its legitimate results; and this may in part explain the necessity for "*eternal secrecy*" relative to the transactions in the confessional.

3. The *restriction* is only made to "carnal sins" (sins against chastity); consequently, the "confessor can grant valid absolution to his accomplice in all other sins; namely, in *theft*, in *homicide*, etc."

Comment here is superfluous. With this system of theology in the hands of professional experts in the arts of seduction and crime, where is the protection for virtue, character, life, or property?

When thousands of the Roman clergy, with their degrading theology, are worming their way into every department of society, instilling in the minds of youth and unsuspecting females the most pernicious sentiments, in the name of religion and by authority of a professedly infallible Church, it ceases to be a matter of surprise that unblushing crime stalks abroad in every department of society. It is not strange that prisons and poor-houses are increasing, that brothels and foundling institutions are multiplying, and that Magdalene institutions have become a necessity in the Church of Rome, and are polluting American soil.

It is not strange, with a system of corruption like this intrenching itself in the midst of pure and virtuous society, that the fond hopes of parents are often blasted, and their gray hairs brought down to the grave sorrowing.

And it is not strange that those who know the facts are willing to jeopard life and property to expose and prevent the evil. The wonder is that this system of clerical debauchery has not been suppressed by enlightened public sentiment or penal enactments.

If the confessional stood alone, it would seem enough to corrupt the nation; but when taken in connection with the system to which it belongs, it is truly the "mystery, Babylon the great, the mother of harlots and and abominations of the earth."

The obligations of secrecy shall not longer conceal the abominations of the confessional. Assumed dignity, or threats of personal violence and assassination, will avail nothing. Deeds of darkness shall be brought to light until legal enactments shall abate the confessional as an intolerable nuisance.

Pope Gregory VII and the Council of Trent prohibited the marriage of the Roman clergy; but the latter permitted the clergy to keep concubines for a money consideration. This fact was clearly established in the celebrated debate between the Rev. Doctors Campbell and Purcell on Romanism. Dr. Campbell affirmed, Dr. Purcell denied, and Dr. Campbell proved that the Roman clergy had kept concubines with the knowledge and approbation of their ecclesiastical superiors. The facts are recorded in the debate, on pages 218, 253, 239, 359, etc. This fact is also confirmed by the "Moral Theology" of St. Liguori, which, for the benefit of the Roman clergy who keep women and profess celibacy, we here insert, translated into English, the following:

"A bishop, however poor he may be, can not appropriate to himself pecuniary fines, without the license of the Apostolic *See;* but he ought to apply them to pious uses. Much less can he apply those fines to any thing else but pious uses which the Council of Trent has laid upon non-resident clergymen, or upon those clergymen who keep concubines." (*Liguori Ep. Doc. Mor.*, p. 444.)

This translation was made by Dr. Campbell, and was not disproved by Archbishop Purcell. The original was found in the edition of Liguori, published A. D. 1832, on the page indicated. We have before us later editions (duplicates), bearing date A. D. 1846. The Latin text is verbatim; but is found in Vol. IX, page 411: "Mulctas pecuniarias Episcopus sibi," etc. Here we have it in plain terms. The Roman clergy are licensed to keep concubines, and the bishops are required to apply to *holy uses* the proceeds of their lustful gratifications. In this, as in other cases, the end sanctifies the means. Truly, with the Roman clergy, money hides a multitude of sins. It will be remembered that the Council of Trent is the last general council but one, and the above license, not being revoked, is now in force. St. Liguori, whose "Moral Theology" contains the facts, is an approved theologian, and a patron saint in the Roman calendar. So that the Roman clergy *now* have the sanction of the Church for keeping concubines, commonly called "*nieces.*" Present circumstances and past history clearly indicate the fact that they often avail themselves of their canonical privileges. It usually requires not less than one girl or woman to keep the priest's house, to provide for his toilet and table They frequently eat, sleep, and live under the same roof, in a state of seclusion. The outside world has only an occasional glance into their secluded devotions; but enough is known to establish the fact that seclusion is not an infallible sign of immaculate purity. High walls, darkened windows, and bolted doors are not inseparably connected with fervent piety. Facts at Evansville, St Wendell's, and Vincennes, Ind; Bardstown, Ky; Alleghany City, Penn.; Montreal, Canada; New York City; Canton, Macomb, and New Berlin, Ill., and at other places too numerous to mention, are sufficient to show that if priests are not licensed by their Church to keep concubines, they ought, for the sake of decency, morality, virtue, religion, and a respect for the laws of God and man, to be lawfully married. The foundling institutions of Italy are a sad comment on the professed sacerdotal celibacy of the clergy, and the vows of chastity taken by priests and nuns.

Those who escape from the tyranny and dungeons of monasteries and convents in the United States and in Rome, bear their united testimony against the chastity of the clergy, and against the purity of monasteries and convents. History shows that, in the *palmy* days of the Roman sect, in the ninth, tenth, eleventh, and twelfth centuries, popes, bishops, and priests were often publicly and notoriously licentious; and the licentious popes are numbered among the infallible successors of St. Peter, through whom the modern clergy profess to obtain power to forgive sins and perform miracles. The late Priest Hogan says, "*Every priest has one concubine, and some have more.*"

When the facilities for concealing vice, and the fearful obligations of secrecy, are understood, it will be a matter of surprise that the secret vices of the clergy are ever detected. It may, for the present, suffice to know that the Roman clergy in our midst live in seclusion, under circumstances which would not be tolerated in any other class of clergymen or professional men.

It might not be amiss for grand jurors to inquire into the domestic relations of the Roman clergy and their pious "nieces," and for State legislatures to open convents to inspection, or require them to stand open, as other institutions. Prison-pens are not needed in which to educate, proselyte, and prostitute American daughters. And until priests and nuns can show a purer record of convents and monasteries in connection with Auricular Confession in Italy, Spain, and other Roman countries, they should at least cease to affect superior sanctity in America.

Canon law requires nuns to confess once each month. Father Garasche, of St. Louis, says that "custom requires them to once a week." Every regular convent should have at least one father-confessor, who may have access to the institution by day or by night.

"Edith O'Gorman," the escaped nun, in her book corroborates these facts, and shows that, when nuns in the school-room and under other circumstances meet Protestant girls whom they are endeavoring to proselyte, they appear as meek as mercy, as pure as love, and as innocent as angels; but in another department, where school-girls are not admitted, and where priests and nuns have intercourse, there is jealousy and strife, and often gross licentiousness; and she defiantly challenged investigation, and pledged herself to prove it. They did not do it; but endeavored to assassinate her, to suppress the facts.

With these facts before the people, how unreasonable that Protestants should persist in patronizing Popery and convents.

CHAPTER IX.

CLERICAL SEDUCTION, HOW CONCEALED.

THE fearful obligations and penalties of Auricular Confession, with the numerous facilities for vice in seclusion, are not considered a sufficient protection for the clergy to prevent disclosures from the confessional. The questions to be propounded in the confessional are adapted to test the virtue of every female who goes there, and thus give the priest unrestrained power to select his victims.

In confession, females are required to confess thoughts, desires, emotions, words, and actions, in detail, and promptly answer all questions propounded by the priest, otherwise not obtain absolution. Having been thoroughly trained to the belief that her salvation is predicated upon thorough confession and implicit obedience to her ecclesiastical superior, she dare not incur his displeasure. The unfortunate victim thus fettered by education, obligations, penalties, and the seclusion of the confessional, may fall a victim to the arts of a skillful seducer. If she willingly consents to his seductive influence, then the matter rests between the parties guilty, who are equally bound by interest and obligation to conceal the facts. If otherwise, and her inherent womanly virtue indignantly resents the lecherous encroachments of her clerical seducer, she has no means of sure redress. She is bound, under *"penalty of damnation,"* to secrecy. A Protestant lady would fly to the strong arm of a father, husband, or brother, and find redress; but she dare not do it; her lips are sealed till she first consults the bishop. If he refuses to interpose his authority, she has no redress whatever. If there is probable danger of scandal, the bishop may at pleasure send the priest to another parish, where he is not so well known, to practice his old tricks. Virtue is not an indispensable qualification for a Roman clergyman. The Council of Trent and moral theology teaches that vice does not annul "holy orders," and that "official acts of a wicked priest are valid." (Catechism of Trent, page 172, etc.)

The fact that seduction is practiced in the confessional can not be successfully denied, and in systems of theology provisions are made by which bishops may so regulate it as to prevent scandal to the Church by suppressing facts. Their method does not remove the evil, but otherwise perpetuates and enhances the pernicious influence. The following interesting instruction to the Roman clergy is found in Dens's Theology, vol. iv, pages 301, 302, 303 :

"DE MODO DENUNTIANDI SOLLICITANTEM PRÆFATUM.	"ON THE MODE OF DENOUNCING THE AFORESAID SEDUCER.
"Primus modus magis conveniens est si ipsa persona sollicitata immediate, nulli, alteri revelando, accedat episcopum	"The first and most convenient mode is this, if the person upon whose chastity the attempt had been made, would

sive ordinarium. 2. Potest episcopo scribere epistolam clausam et signatam sub hac forma: 'Ego Catharina N., habitans Mechlinæ in platea N. sub signo N. hisce declaro me 6 Martii anni 1758 occasione confessionis fuisse solicitatum ad inhonesta a confessario N. N. excipiente confessiones Mechlinæ, in ecclesia N. quod juramento confirmare parata sum.'" (Dens, tom. vi, p. 302.)

proceed herself immediately to the bishop or the ordinary, without revealing the circumstance to any one else. 2. She can write a letter, closed and sealed, to the bishop, in the following form: 'I, Catharine N., dwelling at Mechlin, in the street N., under the sign N., by these declare, that I, on the 6th day of March, 1758, on the occasion of confession, have been seduced to improper acts by the confessor N., hearing confessions at Mechlin, in the church N., which I am ready to confirm on oath.'" (Dens, vol. vi, p. 302.)

"3. Si autem scribere nequeat, similis epistola scribatur ab alio v. g. a secundo confessario cum licentia pœnitentis; et nomen pœnitentis seu personæ sollicitantis, exprimatur ut supra: sed nomen confessarii sollicitantis ut occultum maneat scribenti, non exprimatur, verum a tertio aliquo, rei ignaro, in 'chartula aliqua nomen ejus scribatur sub alio prætexta; quæ chartula epistolæ præfatæ includatur."

"3. But if she can not write, let a similar letter be written by another, namely, by a second confessor, with the license of the penitent, and let the name of the penitent or person seduced be expressed as above, but let the name of the seducing confessor, in order that it may remain a secret to the writer, be not expressed, but let his name be written under a different pretext, by some third person, ignorant of the circumstances, on some scrap of paper, which may be inclosed in the aforesaid letter."

"In hoc casu (denunciatonis) tamen quidam putant moderandum, et considerandas esse

"In this case (of denouncing), however, some are of opinion that moderation must

circumstantias frequentiæ, periculi," etc. (Dens, tom. vi, p. 301.)

"Monentur interea confessarii, ut mulierculis quibuscumque accusantibus priorem confessarium, fidem leviter non adhibeant; sed prius scrutentur accusationis finem et causam examinent earum mores, conversationem," etc. (Dens, vol. vi, p. 302.)

"Quocirca observa, quod quæcumque persona, quæ per se vel per aliam, falso denuntiat sacerdotem tanquam sollicitatorem, incurrat casum reservatum summo Pontifici. Ita Benedictus XIV. Constit. Sacramentum Pœnitent." (Apud Antoine, p. 418.)

"Benedictus XIV, in Constit. citata numero 216, reservavit sibi et successoribus peccatum falsæ denuntiationis confessarii sollicitantis ad turpia, sed sine censura." (Dens, tom. vi, p. 303.

be observed, and that the circumstances of frequency, of danger, etc., must be considered." (Dens, vol. 6, p. 301.)

"In the meantime, confessors are advised not lightly to give credit to any woman whatsoever, accusing their former confessor, but first to search diligently into the end and cause of the occasion, to examine their morals, conversation," etc. (Dens, vol. vi, p. 302.)

"For which reason observe, that whatever person, either by herself, or by another, falsely denounces a priest as a seducer, incurs a case reserved for the supreme pontiff. Thus Benedict XIV, in the constitution, called 'Sacramentum Pænitentiæ." (In Antoine, p. 418.)

"Benedict XIV, in the constitution cited in No. 216, reserves to himself and his successors the sin of falsely denouncing a confessor for seducing his penitent to commit carnal sin." (Dens, vol. 6, p. 303.)

The above extracts might be enlarged; but they are sufficient to illustrate several important facts:

1. That seduction and licentiousness are practiced in the solitude of the confessional, where there is no witness to corroborate the testimony of the outraged and

insulted female, and when the word of the guilty priest would be a sufficient refutation of the charge of guilt. Peter Dens shows in subsequent numbers that the testimony of a priest is to be taken in preference to that of a layman.

2. The fact that a chapter in an approved system of theology is devoted exclusively to the *best* method of denouncing seducing confessors is evidence of the prevalence of the practice. If there be no such thing as a will or a warrantee deed there would be no necessity for a formula for one. If there is no such thing as clerical seduction in the confession, there would be no necessity for chapters of detailed instructions as to the best methods of concealing the facts from the people and making them known to the bishop.

3. Female penitents are not compelled to report the licentiousness of the priests to the bishop; but if they do not report to him, they can not to any other person, except to another priest, who may be as vile as the seducer, and may thereby take occasion to add insult to injury.

Priests are not infallible, and St Liguori being judge, they are not distinguished for virtue. He says: "Among the priests who live in the world, it is rare, and very rare, to find any that are good." Regarding this as a settled fact, the Saint has given rules to confessors to guard them against the influence of their constitutional and habitual weakness. He says:

"The confessor ought to be extremely cautious how he hears the confession of women, and he should particularly bear in mind what is said in the holy congregation of bishops, 21 January, 1610: '*Confessors should not, without necessity, hear the confessions of women after dusk or before twilight.*' In regard to the prudence of a confessor, he ought, in general, rather to be rigid with young women, in the confessional, than bland; neither ought he to allow them to come to him before confession to converse with him; much less should he allow them to kiss his hands. It is also imprudent for the confessor to let his eyes wander after his female penitents, and to gaze upon them as they are retiring from confession. The confessor should never receive presents from his female penitents; and he should be particularly careful not to visit them at their houses, except in case of severe illness; nor should he visit them then, unless he be sent for. In this case he should be very cautious in what manner he hears their confessions; therefore the door should be left open, and he should sit in a place where he can be seen by others, and he should never fix his eyes upon the face of his penitent; especially if they be spiritual persons, in regard to whom, the danger of attraction is greater. The venerable father, Sertorius Capotus, says, that the devil, in order to unite spiritual persons together, always makes use of the pretext of virtue, that, being mutually affected by these virtues, the passion may pass from their virtues over to their persons. Hence, says St. Augustin, according to St. Thomas, confessors, in hearing the confessions of spiritual women, ought to be brief and rigid; neither are they the less to be guarded against on account of their being holy; for the more holy they are, the more they attract.' And, he adds, 'that such persons are not aware that the devil does not, at first, lance his poisoned arrows, but those only which touch but lightly, and thereby increase the affection. Hence, it happens, that such persons do not conduct themselves as they did at first, like angels, but as if they were clothed with flesh. But, on the contrary, they mutually eye one another, and their minds are captivated by the soft and tender expressions which pass between them, and which still seem to them to proceed from the first fervors of their devotion; hence they soon begin to long for each other's company; *and thus*, he concludes, 'the spiritual devotion is converted into carnal. And, indeed, O, *how many priests*, who before were innocent, have, on account of these attractions, which begin in the spirit, lost both God and their soul!' (*Liguori*, N. 119.)

"Moreover, the confessor ought not to be so fond of hearing the confessions of women, as to be induced thereby to refuse to hear the confessions of men. O, how wretched it is to see so many confessors, who spend the greater part of the day in hearing the confessions of certain religious women, who are called *Bizocas* (a kind of secular nun), and when they afterwards see men or married women coming to confession to them, overwhelmed in the cares and troubles of life, and who can hardly spare time to leave their homes, or business, how wretched it is to see these confessors dismiss them, saying, '*I have something else to attend to; go to some other confessor;*" hence it happens, that, not finding any other confessor to whom to confess, they live during months and years without sacraments, and without God." (N. 120.)

It is evident that but little regard is at present given to these instructions. Priests are not deterred by darkness from hearing confessions. The obligations of secrecy, the obscurity of the confessionals, the closed doors, the whispering, and especially the corrupting nature of the communications, all declare that the greater the darkness the more fervent the devotions of the confessor. The great and infallible teacher said, "Men have loved darkness rather than light; because their deeds were evil."

The whole instruction indicates that, while the priest in the confessional is pretending to act the part of God, and forgive sins, he is really a creature of burning lust. It also shows how much more devout the priests are in hearing the confession of women than men, and the confession of nuns and young women than the married. These facts from the pen of a saint in the Church, with the indorsement of popes and bishops, speak in accents of thunder to Protestants, Beware of the confessional, because of the affected superior sanctity of priests and nuns! We, about two years since, heard Father Garesche, of St. Louis, the distinguished Jesuit priest and lecturer, eulogize the nuns, whom he said, "went to confession every week." He said, "Look at them as they walk the streets, as pure as the driven snow." We thought that, with equal propriety, an old Pharisee might have leaned lovingly against a whited sepulcher, and exclaimed, What a beautiful and pure vault is this; no dead men's bones here! Here Father Geresche waxed warm in his affected zeal, and said, in substance, and we believe in exact words, "If any man should say to me that there is any thing obscene in Auricular Confession, clergyman as I am, I would knock his teeth down his throat." On the next day we inserted in the *Bloomington Journal* the following:

"A CHALLENGE.

"FATHER GARESCHE: *Sir*—Since you have appeared before the citizens of Bloomington as a champion of Romanism, which claims for its adherents *exclusive salvation* through 'Auricular Confession' and the 'Power of the Keys,' and since, on last night, you publicly declared and endeavored to prove that God had appointed the Roman clergy 'Vicegerents' and 'Vicars of Jesus Christ,' with power to 'forgive and retain sins,' and since you made a denunciatory effort to conceal the *horrible obscenities* of the confessional, you will permit me respectfully to discard your denunciations, deny your arrogant assumptions, and hereby challenge you to prove that *the doctrine of Auricular Confession as taught in the Roman Catholic Church, is accordant with reason and the Bible.* The discussion to be in the presence of *men only*, under stipulated rules, and at such time and place as shall be mutually adjusted. And, inasmuch as the dogma to be discussed is one of the main pillars of your creed, without which your absolution is a blasphemous, ecclesiastical farce, I shall expect an early and favorably reply. As to your terrible threat that *you* 'would knock a man's teeth down his throat who said that there was any thing obscene in Auricular Confession,' I take the risk, and hereby give you due notice that I not only *say it*, but I am prepared to prove that *horrible obscenity is authorized in the confessional*, and horrible crimes have been sanctioned by it, and that the Roman clergy are authorized to perjure themselves to conceal the abominations. My present address is St. Nicholas Hotel, Bloomington, Ill. Respectfully,

J. G WHITE,
Minister of the Cumberland Presbyterian Church.
JANUARY 8, 1873.

Obtuse as are the perceptions of parish priests, we thought that Father Garesche would understand the above; but he preferred to continue to whisper in the ears of females his *chaste* theology, about "De usu conjugii," "De luxuria," and kindred subjects. We followed him with lectures, and defiantly challenged him to discussion, but our teeth sustained no damage. We desired to meet him in the presence of men alone, where we might, with propriety, quote from his approved theology. The fact is, the subject will not admit thorough investigation, except in presence of men of mature age. How long shall slumbering Protestants wink at the abominations of the confessional, and permit bachelor priests and their licentious theology to corrupt society? Will not the just indignation of outraged society demand legal enactments to suppress the confessional in common with brothels, to which it is so nearly allied, and to which it so largely contributes? Let facts be generally understood and outraged virture will demand the suppression of the confessional and professed sacerdotal celibacy, as prolific sources of licentiousness.

CHAPTER X.

CORRUPTION OF THE CONFESSIONAL.

BEFORE Auricular Confession was established by canon law the Church of Rome was notoriously corrupt. Ambition, usurpation, and avarice exerted their legitimate influence; and under the influence of Auricular Confession she has continued to fall more deeply into the abyss of corrupting error. An examination of the lives of the popes presents a record of crime and scandal scarcely paralleled in the history of pagan Rome, and sufficient to fill every mind with horror and disgust.

Roman historians are compelled to admit the disgraceful corruption of many of the popes through whom they claim an "unbroken, holy, apostolic succession."

A few facts from history may exhibit links in this chain of infallibility. Boniface VIII, Calixtus III, John XIII, and Boniface IX were notoriously covetous. Benedict XII, Adrian IV, Celestine III, Innocent IV, Alexander III, Gregory XIII, Clement V, VI, and VII, Boniface VIII, Paul II, John XXIII, and numerous other popes, were proud as Satan, which is but one characteristic of all the popes. Sylvester III, and all his successors for nine or ten popes, were professed conjurers. Licentiousness has been a distinguishing characteristic of the popes—their number too numerous, and their crimes too abominable to mention.

John XII, Gregory V, John XIII, Boniface VII, Benedict IX, Innocent III, were murderers. Many popes were instigators of jealousies and discord which cost the lives of thousands.

Several popes have been schismatics, two and three contending for the supremacy at the same time. These schisms varied from two to six, seven, thirteen, sixteen, twenty, and to thirty-nine years; and, during these periods of ecclesiastical strife, popes cursed each other and fought against each other, while multitudes were slaughtered by their cruel ambition. The base and licentious popes, John X, XI, and XII, were golden links in the apostolic chain of Romish succession. And, if supremacy necessarily implies infallibility, the *prostitutes*, Marosia, Theodora, and the Countess of Tuscany, might each set up a plausible claim, as they held at pleasure, for a time, the patrimony of St. Peter.

Barronius, the great Roman historian, declares that Pope John XI was "*a monster of iniquity:*" that Pope John XII was "a gambler, a whoremonger, seducer, Sabbath-breaker, bloodthirsty, and a man capable of all iniquities, and that he died in the midst of debauchery."

These facts are attested by the writings of numerous Roman historians.

St. Liguori admits that "many priests have lost both God and their own souls, by hearing the confession of women, and holding communications with them."

The following facts, from the pen of Rev. Joseph Reeve, a distinguished historian of the Roman Church, may exhibit a few links of the holy, unbroken succession through which the Roman clergy profess to receive power to forgive sins, and from whose sacred treasure of surplus righteousness they dispense indulgences. He says:

"Italy, from the end of the ninth century, as we have seen, was become the seat of faction and civil discord. The ecclesiastical state was kept in a long and disgraceful servitude by the ambition of rival senators, by the Marquises of Tuscany, and the Earls of Tusculum. By these petty tyrants, the patrimony of St. Peter was torn to pieces, and sacrilegiously usurped. The popes were not masters of their own capital. Raised by faction, as it happened, or by intrigue, they lost their personal respectability, were often insulted, imprisoned, and even murdered, by the prevailing party.

"Two sisters, prostitutes, Marozia and Theodora, daughters of the lewd Marchioness of Tuscany, governed Rome by their political influence and criminal intrigues. To these disorders the popes themselves contributed in no small degree. After Stephen IV, who died in 891, succeeded Formosus, Stephen VII, Romanus, Theodore II, John IX, Benedict IV, Leo V, Christopherus, Sergis III, Anastasious III, Lando, John X, Leo VI, Stephen VIII, John XI, Leo VII, Stephen IX, Martinus II, Agapitus II, John XII, Leo VIII, Benedict V, John XIII, Benedict VI, Donus II, Benedict VII, John XIV, Boniface VII, John XV, Gregory V, Sylvester II.

"Between the years 891 and 999, here are one and thirty popes. Their number is a clear proof that the reign of many of them was short, and their end dishonorable. Sergis III exhibited a spectacle of scandal of which the Christian world had never known an example, a sovereign pontiff clasped in the embraces of a notorious prostitute. Sergis III, without regard for the dignity or the holiness of his pontifical character, publicly avowed his criminal connection with Marozia. By her he had a son, who, under his mother's influence, crept afterward into St. Peter's chair, by the name of John XI. To the infamy of his spurious birth he added personal vice, in which he was shamefully imitated by many who, in that century, were raised to the papal throne without the virtues to merit or support their elevation." (Reeve's Ch. Hist., p. 291.)

Here are specimens of Romish infallibility. Here are works of supererogation, with a vengeance. A viler set of whoremongers and drunkards were, probably, never congregated this side of hell; and yet these drunken, debauched, shameless beasts, in the form of men, were at the head of the Romish sect, which professed infallibility, and pretended to perform superabundant good works.

Again, Reeve says, pages 315, 316:

"Simony and incontinence had struck deep root among the clergy of England, Italy, Germany, and France. [A. D. 1074.] The evil began under those unworthy popes who so shamefully disgraced the tiara by their immoral conduct in the tenth century. The scandal spread, and had now continued so long that the inferior clergy pleaded custom for their irregularities. Many of the bishops were equally unfaithful to their vows, and with greater guilt. Hence the corrupt laity, being under no apprehension of reproof from men as deeply immersed in vice as they, gave free scope to their passions."

Here is a specimen of Romish virtue and holiness, when popes were in power, when kings and emperors bowed their necks to ecclesiastical dictators, and *vile prostitutes* and their pontifical paramours reveled in gross licentiousness.

These were the palmy days of papal Rome, the sacred memory of which she now cherishes, and for the return of which her energies are concentrated, and her priests and nuns pray most devoutly. These were the triumphal days of the mother of harlots. May they never return to curse the world!

In that darkest period of the world's history, when licentiousness overwhelmed the Romish sect, including bishops and popes, the doctrine of Auricular Confession, and other kindred heresies, were projected by the clergy, and accepted by ignorant dupes of papal despotism.

Truth promotes intelligence and virtue. Auricular Confession is the offspring of Popery, and a prolific source of ignorance and vice.

The Bible teaches that God alone can forgive sins. The doctrine of Auricular Confession is predicated on the blasphemous assumption that any priest, bishop, or pope can forgive sins.

God proclaims pardon on the condition of faith, without money or price.

The Roman clergy offer pardon on condition of penance with both money and price.

God declares that he blots out transgressions, and re-remembers iniquities no more.

The Romish clergy say sins may be pardoned, but must afterward be purged in purgatory, if not commuted by indulgences.

The Lord said, "Cursed be the man that trusteth in man, and maketh flesh his arm."

The Romish clergy say the man shall be blessed who trusteth in *them* and the merits of saints.

The Lord said:

"Thou shalt have no other gods before me.

"Thou shalt not make unto thee any graven image, or the likeness of *of any thing* that *is* in heaven above, or that *is* in the earth beneath, or that *is* in the water under the earth:

"Thou shalt not bow down thyself to them, nor serve them." (Exodus xx, 3–5.)

The Roman clergy say you *shall* make images and pictures, and bow down before them; you shall worship Gods made of bread, wine, and wax; you shall wear scapulars, repeat prayers of beads, kiss medals, and above all, pray devoutly to the Virgin Mary, and by so doing, you will obtain indulgences, escape purgatory, contribute to the *treasury* of the Church, and be infallibly sure of heaven. Thus the Word of God and the Romish clergy contradict each other.

Let God be true, and the doctrine of Auricular Confession a lie.

> "A strange belief that leaned its idiot back
>   On folly's topmost twig—
> A lazy, corpulent,
>   And ever credulous faith, that
> Stepped on, but never earnestly inquired
>   Whether to heaven or hell the journey led."

CHAPTER XI.

CORRUPTION OF THE CONFESSIONAL, CONTINUED.

REV. JOSEPH BLANCO WHITE, once a distinguished priest in Spain, and subsequently a clergyman of the Church of England, bears his testimony to the gross and revolting licentiousness of the Spanish clergy. (See Practical and Internal Evidences against Catholicism.)

This work, of two hundred and eighty-four pages, discloses the inner life of priests and nuns in convents and monasteries, and the corrupting influence of the confessional on the minds of youth and females. It also exhibits, to an alarming extent, the power of the confessional to enslave men, and sustain a corrupt ecclesiastical despotism. The author exposes the criminal character of the clergy of Spain, and declares that, in their bacchanalian orgies, they often, with jest and ribaldry, disclose the obscene communications of the confessional, and boast of the number of children born unto them within a *"few days"* by their concubines. On account of the indelicacy of the subject, he acknowledged his inability to present the facts in a work published for the general reader.

On pages 111 and 112, he says of the priests:

"I have known the best among them; I have heard their confessions; I have heard the confessions of young persons of both sexes, who fell under the influence of their suggestions and example; and I do declare that nothing can be more dangerous to youthful virtue than their company.

"How many souls would have been saved from crime, but for their vain display of pretended superior virtue, which Rome demands of her clergy.

"The cares of married life, it is said, interfere with the duties of the clergy. Do not the cares of a vicious life, the anxieties of stolen love, the contrivances of adulterous intercourse, the pains, the jealousies, the remorse, to a conduct in perfect contradiction with a public and solemn profession of virtue—do not these cares, these bitter feelings, interfere with the duties of the priesthood?"

Thus writes a man of culture, who once stood high in the estimation of the Roman clergy, and for many years had personal knowledge of the practical results of Auricular Confession. He draws aside the curtain, and discloses the hidden mysteries of convent life in relation to the clergy and the confessional.

He says, on page 112:

"The picture of female convents requires a more delicate pencil; yet I can not find tints sufficiently dark and gloomy to portray the miseries which I have witnessed in their inmates. Crime, indeed, makes its way into those recesses, in spite of the spiked walls and prison gates which protect the inhabitants. This I know, with all the certainty which the self-accusation of guilt can give."

He further shows, what ought to be known to every American citizen, that nuns are slaves to their ecclesiastical superiors. That the Council of Trent enjoins on all bishops to enforce the close confinement of nuns by every means, and even to engage the assistance of the secular arm for that purpose; and entreats all princes to protect the inclosure of the convents; and threatens excommunication on all civil magistrates who withhold their aid when the bishops call for it.

The Council of Trent says:

"Let no professed nun come out of her monastery under any pretext whatever, not even for a moment."

"If any of the regulars [men or women under perpetual vows] pretend that fear or force compelled them to enter the cloister, or that the profession took place before the appointed age, let them not be heard, except within five years of their profession. But if they put off the frock of their own accord, no allegation of such should be heard: but, being compelled to return to the convent, *they must be punished as apostates*, being, in the mean time, deprived of all the privileges of their order."

In the cases of Milly M'Pherson, of Kentucky, and Barbara Rubrick, of Craco, Austria, we have illustrations of the Christian punishment of nuns who violate convent rules, or refuse to submit to the dictation and seduction of their father confessors. It is evident that the former went to a premature grave by the hand of an assassin, or to a prison from which she has not escaped, and the latter to a dungeon twenty-one years, and after suffering, mentally and physically, more than a thousand deaths, she was rescued by the civil authorities, and survived long enough to recite her story of woe. According to their statements they were punished because they refused to prostitute themselves in complyance with the wishes of licentious priests. The records of the court at Bardstown, Kentucky, disclose the facts in the case of Milly M'Pherson. Rev. N. L.

Rice published the results of a malicious prosecution on the part of priest Elder and others, which was only the dodge of a Jesuit cuttle-fish to divert attention from the guilty party, and save the character of the convent. The report was of a nature to damage an innocent priest and a pure institution of learning. In this case the jury estimated the damage at "*one cent*," and doubtless they took into the account the fact that some men and institutions are not susceptible of being slandered, their corruption is so notorious.

The acknowledged ability and veracity of Dr. Rice gives weight to all he says and does, and the fact of only "*one cent*" damage assessed, and under special instruction, is conclusive evidence that the jury understood the facts, and, as they certify, would have returned a verdict for the defendant but for the instruction of the Judge. The damaging fact is, that more than thirty-five years have passed away, and Milly M'Pherson has not been found. The fact that the Roman clergy did not produce her in court in order to gain their ten thousand dollars, is evidence conclusive that they could not. Men who perform fœtal baptisms and sell indulgences, would not knowingly neglect such a *golden* harvest.

Doubtless Milly M'Pherson has long since gone forever beyond the corrupting influences of the confessional and convent.

The Italian patriot Gavazzi says:

"The latest efforts of confessors are against civil and religious freedom. . . . Remember my words, and may they be profitable! We have in Italy (and their mystical operation extends all over the world) three bulls, of three different popes, Pius VII, Leo XII, and Gregory XVI, obliging all penitents to discover all among their relatives who are adherents to the liberal cause. Thus all names of all patriots are known to the authorities of the Church; so that in my Italy such a control over *one* heart (generally a female one) implicates many in the mesh. Sisters betray their brothers, wives their husbands, and —what is horrible to say, what is against the law of nature, what is possible only in the cruel system of Rome—*mothers* are obliged to accuse their poor children! We have in Italy not one, but hundreds of thousands of brothers, husbands, and sons, young men, condemned to the galleys, exile, the scaffold, only in order that their sisters, wives, and mothers can receive sacramental absolution from the priests. You will perhaps say: that does not touch us—such kind of perfidy never will approach American shores; Americans, Americans, you mistake Popery! Here she must be in disguise; but in her heart she is always Popery. And secretly she will do in America what openly in Italy. She can not be better in her nature because Americanized. If you do not know the system, hear for your benefit what it is abroad, in order to save from its snares your dear country. In the short but glorious period of our Roman Republic (Americans! hear an Italian), we found in the palace of the Inquisition at Rome a large book, with the correspondence of all the bishops throughout the Christian world, in which correspondence we found the names of all patriots, leaders of liberals, among all nations, not only Italians, but Frenchmen, Spaniards, Portuguese, AMERICANS, Mexicans, all diligently collected, and sent from the bishops of all the Christian world to the Inquisition of Rome. This is Auricular Confession!" (Gavazzi's Lectures, pages 242, 243, and 244.)

Those who have not read Gavazzi's lectures would do well to read them with special reference to the confessional and the corrupting influence of the clergy through the confessional.

TESTIMONY OF AN EYE-WITNESS.

The Rev. L. Giustiniani, D. D., once a priest in Rome, afterward a minister of the Evangelical Lutheran Church, published a work in 1843, entitled *"Papal Rome as it Is. By a Roman."* It is worthy of being read by every American citizen. The Doctor having resided in the city of Rome itself, the very "seat of the beast," and who was therefore perfectly acquainted with the practical operation of Auricular Confession, gives numerous illustrations of its corrupting influence from personal observation, two of which we here present.

The first is in reference to a young lady of about seventeen years old, in the family where the Doctor was boarding. He says:

"One day the mother told her daughter to prepare to go with her to-morrow to confess and to commune. The mother, unfortunately, feeling unwell the next morning, the young lady had to go by herself; when she returned, her eyes showed that she had wept, and her countenance indicated that something unusual had happened. The mother, as a matter of course, inquired the cause, but she wept bitterly, and said she was ashamed to tell it. Then the mother insisted; so the daughter told her that the parish priest, to whom she constantly confessed, asked her questions this time which she could not repeat without a blush. She, however, repeated some of them, which were of the most licentious and corrupting tendency—which were better suited to the lowest sink of debauchery than the confessional. Then he gave her some instructions which decency forbids me to repeat; gave her absolution, and told her, before she communed, she must come into his house, which was contiguous to the church; the unsuspecting young creature did as the father confessor told her. The rest, the reader can imagine. The parents, furious, would immediately have gone to the archbishop, and laid before him the complaint; but I advised them to let it be as it was, because they would injure the character of their daughter more than the priest. All the punishment he would have received, is a suspension for a month or two, and then be placed in another parish, or even remain where he is. With such brutal acts, the history of the confessional is full." (Papal Rome as it Is, pages 83 and 84.)

The other illustration is in reference to the manner of confessing sick penitents in their bed chambers in the city of Rome. On pages 188 and 189 the Doctor says:

"Only go to Rome and you will see the indisposed fair penitent remain in her bed, and the Franciscan friar, leaving his sandals *before* the door of her bed-chamber, as an indication that he is performing some ecclesiastical act, then *none*, not even the *husband*, can enter the chamber of his wife, until the Franciscan friar has finished his business and leaves the chamber; then the husband, with reverence, ready waiting at the door, kisses the hand of the father Franciscan for his kindness for having administered *spiritual comfort* to his wife, and very often he gives him a dollar to say a mass for his indisposed spouse. . . .

"But why shall I speak of the moral corruption of Popery in Rome? It is everywhere the same; it appears differently, but never changes its character. In America, where female virtue is the characteristic of the nation—the only stronghold of the American Republic—it is under the *control* of the priest. If a Roman Catholic lady, the wife of a free American, should choose to have the priest in her bed-room, she has only to pretend to be indisposed, and asking for the spiritual father, the confessor, no other person, not even the husband, dare enter. In Rome it would be at the risk of his life; in America, at the risk of being excommunicated, and deprived of all spiritual privileges of the Church, and even excluded from heaven."

A celebrated orator in the Council of Trent, *Father Antonius Pagannes*, is reported to have said:

"I am silent respecting adulteries, rapes, and robberies; I pass over the great effusion of Christian blood, unlawful exactions, impositions gratuitously accumulated, and for whatever cause they were introduced, persevered in without cause, and the innumerable oppressions of this kind; I pass over the proud pomp of clothing, extraordinary expenses beyond the requirements of their station in life, drunkenness, surfeits, and the inordinate filthiness of luxury, such as never took place before. Women were never less modest and bashful; young men were never more unbridled and undisciplined; the old were never more irreligious and foolish; in fine, never was there in any person less fear of God, honor, virtue, and modesty; and never did carnal licentiousness, abuse, and irregularity prevail to such an extent. For what greater abuse and irregularity can be imagined than a pastor without watchfulness, a preacher without works, a judge without equity, a lawyer without honesty, a magistrate without decorum, laws without observance, a people without obedience, religious professors without devotion, the rich without shame, the poor without humility, women without purity, the young without discipline, the old without prudence, *and every Christian without religion!*"

Such is the picture of unrestrained Popery, as drawn by one of its friends at the Council of Trent, and such are its characteristic influences every-where, and in all ages.

Rev. John Dowling, D. D., a distinguished Baptist clergyman of New York, writing on Auricular Confession, says:

"A single fact will be sufficient to show the awful extent, in Popish countries, of this crime of illicit intercourse with females at confession. About 1560, a bull was issued by Pope Pius IV, directing the Inquisition to inquire into the prevalence of this crime, which begins as follows: 'Whereas, certain ecclesiastics in the kingdom of Spain, and in the cities and dioceses thereof, having the cure of souls, or exercising such cure for others, or otherwise deputed to hear the confessions of penitents, have broken out into such heinous acts of iniquity as to abuse the sacrament of penance in the very act of hearing the confessions, nor fearing to injure the same sacrament, and him who instituted it, our Lord God and Savior Jesus Christ, by *enticing and provoking, or trying to entice and provoke, females to lewd actions, at the very time when they were making their confessions,*' etc.

"Upon the publication of this bull in Spain, the Inquisition issued an edict requiring all females who had been thus abused by the priests at the confessional, and all who were privy to such acts, to give information, within thirty days, to the holy tribunal; and very heavy censures were attached to those who should neglect or despise this injunction. When this edict was first published, such a considerable number of females went to the palace of the inquisitor, in the single city of Seville, to reveal the conduct of their infamous confessors, that twenty notaries, and as many inquisitors, were appointed to minute down their several informations against them; but these being found insufficient to receive the depositions of so many witnesses, and the inquisitors being thus overwhelmed, as it were, with the pressure of such affairs, thirty days more were allowed for taking the accusations; and this lapse of time also proving inadequate to the intended purpose, a similar period was granted not only for a third but a fourth time. Maids and matrons, of every rank and station, crowded to the Iniquisition. Modesty, shame, and a desire of concealing the facts from their husbands, induced many to go veiled. But the multitude of depositions, and the odium which the discovery threw on Auricular Confession and the Popish priesthood, caused the Inquisition to quash the prosecutions, and to consign the depositions to oblivion." (See Dowling's History of Romanism, pp. 335, 336.)

We extract the following from pages 166, 167, of "Romanism not Christianity," by Rev. N. L. Rice, D. D., a distinguished minister of the Presbyterian

Church, whose intelligence and veracity may not be questioned:

"I beg leave here, also, to adduce the testimony of Waddy Thompson, Esq, late Minister Plenipotentiary of the United States at Mexico, concerning the character of the clergy of that country. He is a gentleman of intelligence and standing, not a member, I believe, of any Church, and not chargeable, so far as I know, with any prejudice against the Roman clergy. He says:

"'I do not think that the clergy of Mexico, with very few exceptions, are men of as much learning as the Catholic clergy generally in other countries. The lower orders of priests and friars are generally entirely uneducated, and, I regret to add, as generally licentious. There is no night in the year that the most revolting spectacles of vice and immorality on the part of the priests and friars, are not to be seen in the streets of Mexico. I have never seen any class of men who so generally have such a *roue* appearance as the priests and friars whom one constantly meets in the streets. Of the higher orders and more respectable members of the priesthood, I can not speak with the same confidence; if they are vicious, they are not publicly and indecently so. Very many of them have several nephews and nieces in their houses, or, at least, those who *call them uncle*. The reason given for the injunction of celibacy, that those who are dedicated to the priesthood should not be encumbered with the care of a family, is, I think, in Mexico, much more theoretical than practical.'"

Dr. Rice adds:

"Such is the character of the priesthood in Roman countries; and I have in preceding lectures proved, even by Roman writers, that in former times even the popes were far more immoral than Thompson represents the clergy of Mexico. Are those of the United States much better? There is a public sentiment in our Protestant country that compels them to walk circumspectly; but the facilities for secret vice, afforded by the confessional and nunneries, are such that they can not be easily detected. Many of them, moreover, are foreigners, whose characters have been formed in Roman countries, where the clergy are generally of loose morals; and they certainly have the appearance, generally, of men not given to a great deal of abstinence—men who give no evidence of extraordinary sanctity."

Will any sane man pretend that Rev. Joseph Blanco White, Rev. N. L. Rice, D. D., ex-Minister Thompson, the Italian statesman Gavazzi, and others, are narrow-minded bigots or illiterate men? It is not enough for the Romish clergy to fall back upon their *assumed* dignity, and pronounce all a Protestant slander. The facts are before the world, and they challenge investigation.

Rev. Edward Beecher says:

"I ask your particular attention to the pernicious influence of Romanism on the morals of community in this respect, that you may learn to what a depth of immorality and vice this country would be plunged if Popery should prevail. By the returns laid before Parliament, it appears that in London the proportion of illegitimate births is four per cent; in Paris, it is thirty-three; in Brussels, thirty-six; in Munich, twenty-five; in Vienna, fifty-one per cent. The amount of immorality thus manifested is a hundred-fold greater in some Romish parts of Europe than in any part of Protestant England. In Rome, the city of popes, cardinals, archbishops, bishops, priests, monks, and nuns, they dare not make returns. But one fact speaks for itself. The number of births in Rome, by Dr. Bowring's returns, is four thousand three hundred and odd per annum; and, by the returns of Mittermeyer, the number of foundlings in the different foundling institutions in Rome, during a period of ten years, gives a return of three thousand one hundred and sixty per annum. Hobart Seymour, from whom I take these statistics, says: 'All this certainly speaks very strongly of the immorality of Rome, or declares that if the mothers be married mothers they are the most unnatural mothers in the world.'

"An examination by Mr. Seymour of the official and

Governmental returns of every Roman Catholic country in Europe, in fifteen or twenty folio volumes, enabled him to say that Popery is universally the mother of vice and crime. Thus, in England the ratio of murders, during ten years, was four to a million; during the same time, in Ireland, it was forty-five to a million per annum, and in the most favorable years never less than nineteen to a million. In Belgium, one of 'the best Romish countries, the murders are eighteen to a million; in France, thirty-one to a million; in Austria, the great pillar of Popery, thirty-six to a million; in Bavaria, a Romish state, including homicides, sixty-eight persons to a million—excluding homicides, thirty to a million; in Italy, in the Venetian and Milanese provinces, forty-five to a million; in Tuscany, forty-two to a million; in the States of the Pope, one hundred to a million; in Sicily, ninety to a million; in Naples, doubly cursed by Popery and the most unmitigable Popish civil despotism, two hundred to a million. The average of all these Papal nations is seventy-five to a million. 'In Italy,' says Seymour, 'the land of popes, cardinals, bishops, priests, monks, and nuns, there is perpetrated such an amount of murder that the number of persons killed every year in cold blood is greater than the number of men that have fallen in some of the most terrific struggles on the modern battle-fields of Europe." (Papal Conspiracy Exposed, by Beecher.)

According to the above returns laid before the British Parliament, about three-fourths of all the children born in *Holy Rome*—the mother of harlots—are *foundlings*. Their mothers are not known, much less their fathers. In this Protestant country, where the people are not yet prepared for a national tax to sustain *"foundling institutions,"* the nondescripts may find a home with the "sisters" and "fathers," where, if history be true, many of them *legitimately* belong.

These are but specimens, and are corroborated by all

unprejudiced minds who have thoroughly investigated the subject. On account of the apathy or criminal in-indifference on this subject of many Protestants in America, we give additional facts.

William Hogan, Esq., formerly a Roman priest, from Ireland, published "A Synopsis of Popery as it was and as it is," in 1849, at Hartford, Connecticut. He certifies that he left Ireland, and ultimately the Romish Church, on account of the corruption of the clergy relative to females in the confessional and the convent. He discloses seduction, infanticide, and murder, in their most horrid forms. He gives times, places, and circumstances, and which, if not true, would have subjected him to the severest penalties; and the fact that he was not prosecuted for slander by the priests implicated, is evidence that they dare not attempt to disprove the facts. He speaks of the confessional, and denominates it a "*woman-trap*," and " nunneries"—modestly called—"are Popish brothels." This language is strong, but fully sustained by the horrible disclosures he makes. The work, to be appreciated, should be carefully read; and, since it can only be had by few, we will quote from it more fully.

When discussing Auricular Confession and Popish nunneries, he says:

"Every crime, as I have stated before, which the Romish Church sanctions, and almost all the immoralities of its members, either originate in, or have some connection with, Auricular Confession." (p. 15.)

Among the influences which cause him to renounce Romanism he enumerates the ruin and early death of innocent and virtuous girls through the confessional and convent. Many of the daughters of Protestant parents, proselyted through the female Jesuits of the nunnery, seduced by their confessors, and poisoned or otherwise disposed of to prevent scandal. He speaks from personal observation and from a knowledge of the facts through the confessional. While he was a priest and had knowledge of the facts, his lips were closed by his obligation of eternal secrecy. He solemnly warns American Protestant mothers against the folly and danger of intrusting their innocent daughters to the seductive influence of the "semi-reverend crones called nuns." He says:

"Every confessor has a concubine, and there are very few of them who have not several. Every nun has a confessor. . . . There is scarcely one of them who has not *been herself debauched by her confessor.*"

This is strong language, but the man who used it had officiated as a priest in the confessional on both continents, and had himself often heard the confession of nuns. Again, on page 28, Mr. Hogan says:

"The mother abbess, or superior of the convent, who invariably is the deepest in sin of the whole, and who, from her age and long practice, is almost constitutionally a hypocrite, appears in public the most *meek*, the most bland, the most courteous, and the most *humble* Christian. She is peculiarly attentive to those who have any money in their own right; she tells them they are beautiful, fascinating, that they look like angels, that this world is not a fit residence for them, that they are too good for it, that they ought to become nuns in order to fit them for a higher and better station in heaven. Nothing more is necessary than to become a Roman Catholic and go to confession. Such is the apparent happiness, cheerfulness, and unalloyed beatitudes of the nuns, that strangers are pleased with them. They invariably make a favorable impression on the minds of their visitors."

Mr. Hogan, after comparing a Roman priest to the anaconda, foul, filthy, and ugly, and when he is hungry seizes upon an object which he intends to destroy, he takes it with him to his place of retreat, and there, unseen, covers his prey with slime, and then devours it. He then adds, on page 32:

"I now declare, most solemnly and sincerely, that after living twenty-five years in full communion with the Roman Catholic Church, and officiating as a Romish priest, hearing confessions, and confessing myself, I know not another reptile in all animal nature so filthy, so much to be shunned, and loathed, and dreaded by females, both married and single, as a Roman Catholic priest, or bishop, who practices the degrading and demoralizing office of *Auricular Confession*."

Again, Mr. Hogan refers to his labors in Albany, New York, where he was kindly received and liberally compensated for his labors by the Church, and where he was also Chaplain of the Legislature, and gives his personal experience in the confessional on pages 46, 47, as follows:

"The Roman Catholics of Albany had, during about two years previous to my arrival among them, three Irish priests alternately with them, occasionally preaching, but always hearing confessions. I know the names of these men; one of them is dead, the other two living, and now in full communion in the Roman Church, still saying mass and hearing confessions. As soon as I got settled in Albany, I had, of course, to attend to the duty of *Auricular Confession*, and in less than two months found that those three priests, during the time they were there, were the fathers of between sixty and one hundred children, besides having debauched many who had left the place previous to their confinement. Many of these children were by married women, who were among the most zealous supporters of those vagabond priests, and whose brothers and relatives were ready to wade, if necessary, knee deep in blood for the holy, *immaculate, infallible Church of Rome.* There is a circumstance connected with this, that renders the conduct of these priests almost frightfully attrocious. There are in many of the Roman Catholic churches things, as Michelet properly calls them, like sentry-boxes, called confessionals. These are generally situated in the body of the church, and priests hear confessions in them, though the priest and lady penitent are only separated by a sliding board, which can be moved in any direction the confessor pleases, leaving him and the penitent ear to ear, breath to breath, eye to eye, and lip to lip, if he pleases. There were none of these in the Romish church of Albany, and those priests had to hear confession in the *sacristy* of the church. This is a small room back of the altar, in which the eucharist, containing, according to the Romish belief, the real body and blood of Jesus Christ is kept while mass is not celebrating in the chapel. This room is always fastened by a lock and key of best workmanship, and the key kept by the priest day and night. This sacristy, containing the wafer which the priests blasphemously adore, was used by them as a place to hear confessions, and here they committed habitually those acts of immorality and crime of which I have spoken.

"These details must be unpleasant to the reader, but not more so than they are to me. I see not, however, any other mode in which I can give Americans any thing like a correct idea of that state of society which must be expected in this country, should the period ever arrive when Popery and Popish priests shall be in the ascendant."

These disgusting details are only admissible in consideration of the fact that the evil exists in our midst; and Protestants are unconscious of danger, and are thereby jeoparding their morality, virtue, and eternal salvation, to gratify the avarice and lust of those who are the sworn enemies of civil and religious liberty, and who would esteem it a religious duty, if in their power, to consign them and their children to the dungeon or the Inquisition.

Mrs. Henrietta Curacciolo, an ex-Benedictine nun, has recently published an octavo volume, of nearly five hundred pages, entitled, "The Mysteries of Neapolitan Convents," with an Introduction by Rev. John Dowling, of New York. It fully confirms the statement of others relative to the corrupting influence of the confessional and convent life. It discloses the interesting fact that, about the year 1860, the Italian Government passed a law suppressing convents and monastic institutions, and precluding the possibility of making any more monks and nuns in the Kingdom of Italy. Much of the property of monks and nuns has been confiscated, and they are allowed a daily stipend for their subsistence during life. When these die, there will be no more monasticism in what then comprised the Kingdom of Italy. It is said that twelve thousand monks, and a large company of nuns, are emigrating, chiefly to South America and the Southern States of North America. The leprous spots of Italy are developing themselves on the sacred soil of America. They are more to be shunned than cholera, palsy, plague, and fever. They rot out the vitals of every nation permanently infected with them.

Mrs. Edith Auffray, wife of Professor Auffray, is striking damaging blows at the corruptions of Popery by her "Convent Life Exposed;" also, her telling lectures to ladies and gentlemen, and her *secret lecture* to ladies alone. She speaks from personal experience, having been educated in the convent near Madison, New Jersey, and having taken the veil as a nun, and subsequently presided as lady superior over the convent in Hudson City, New York, from which she declares she fled for her virtue and her life.

Her book shows that her confessor exhausted his arts of attempted seduction, while she in vain appealed for protection to her ecclesiastical superiors. She is probably one of the most persecuted women on the Continent of America.

Mrs. Auffray is better known by her maiden name, "Edith O'Gorman." She united with the North Baptist Church, in Jersey City, New Jersey, and was married to Professor Auffray by the pastor of that Church.

Her published statements and (as we are informed) her private lecture to ladies corroborate the testimony of others, and especially the secret ritual and theology of the clergy. The united testimony of canon laws, catechisms, manuals, rituals, the theology of the Church, the history of individuals, cities, and nations, in connection with the Bible, combine to show that Auricular Confession is unscriptural, unholy, unreasonable, and an unauthorized blasphemous usurpation. It degrades and enslaves the people, subserves the sordid interest and lust of ecclesiastical tyrants, destroys domestic happiness, and is a blighting curse to the social, civil, and religious institutions of any people, city, or nation, in proportion to its unrestrained influence. Such a pestilential evil merits the everlasting execration of all right-minded persons, and the enactment of penal laws for its immediate and perpetual suppression.

CHAPTER XII.

CORRUPTION OF THE CONFESSIONAL, CONTINUED.

THE character of men and systems of religion may be determined by their influence and results. The great and only infallible Teacher said: "By their fruits ye shall know them." This test applies to Romanism with damaging results. After a careful examination of the preceding pages, it will not be difficult to comprehend the fact that Romanism rests as an incubus on all departments of society, and that crime and poverty are its natural and legitimate results.

The judicial statistics of Ireland show that larger numbers of the constabulary are required, in proportion to the population, in Roman Catholic than in Protestant counties. The police statistics are from the census of 1861.

The population of the County Antrim is 247,564; the population of Tipperary is about the same number, 249,106. But while 272 policemen are sufficient to preserve the peace in Antrim, 1,122, or more than four times the number, are required to keep the peace in Tipperary. Nearly the same disproportion prevails in other counties. Down has but 276 policemen, while Galway, with a smaller population, has 691. Westmeath, with a population of 90,879, requires 298 constables, while Londonderry, with a population of 184,209, has but 152. Armagh has 33,000 more people than Roscommon, but while the northern county is kept in order by 193 constables, the western county requires 410. The *Belfast News Letter* justly ascribes this difference to RELIGION, as wherever the Roman Catholics predominate, there the police force is large and costly; but in every county which has a Protestant majority of inhabitants, the constabulary force is small and has little to do. Even in the distinctively Protestant counties, Roman Catholic criminals are in the majority. Thus, while Roman Catholics are less than one-third of the population of the County Antrim, they supply a larger number of prisoners than the Protestant two-thirds. The contrast is still greater in Londonderry and Fermanagh. The Protestants of Ireland bear to Roman Catholics the proportion of 13 to 45. But Protestant prisoners committed in 1863 bore to the Roman Catholics the proportion of 6 to 45, the total number being 4,391 Protestants, against 29,263 Roman Catholics.

The *expenses* of Popery to the country, through its *immorality and crime*, has been used as an argument with such telling effect against the *policy* of Popish endowments, that the demagogues among the Papists have been obliged to resort to measures the most ignoble to blunt its force. They have, by their representatives in the House of Commons, reported crimes against the Protestants which were never committed; and they have perpetrated crimes *against themselves*, in order to charge them upon their heretic neighbors. Galway, from 1851 to 1861, rose, through the operations of Protestant missions, from the twenty-fourth to the seventh place in the scale of morality. Down sank, through Jesuitical intrigues, from the first to the fifth place—*convictions for crime* being the basis.

THE CRUELTY OF ROMANISM.

1 Protestant and 13 Popish Countries.	Population.	Average Murders in a Year.	Yearly murders in one mill'n.
England and Wales	17,927,609	72	4
Belgium	4,347,673	84	19
Ireland, 1851	6,533,579	130	19
Sardinia	4,916,081	101	24
France	35,400,486	1089	30
Austria	36,514,466	1325	36
Lombardy	5,047,472	225	44
Tuscany	1,489,000	84	56
Bavaria	4,520,751	311	68
Malta	114,000	114	80
Sicily	1,936,033	174	90
Papal States	2,908,115	339	113
Naples	6,066,900	1045	172
Spain	12,386,841	——	250

THE IMMORALITY OF ROMANISM.

1 Protestant and 10 Romish Towns.	Legitimate Births.	Illegitim. Births.	Illegitim. B'ths to 100 Legit. Bs.
London, 1851	76,097	3,203	4
Paris, 1851	21,698	10,685	49
Peripignan, 1845	512	256	50
Brussels, 1850	3,448	1,835	53
Tours, 1845	582	330	56
Munich, 1845	1,726	1,702	98
Rodez, 1845	233	243	104
Geuret, 1845	97	115	108
Vienna, 1849	8,881	10,360	116
Mont de Marsan, 1845	74	173	179
Rome, 1845	1,213	3,160	260

The above startling facts are corroborated by additional figures from both continents. The official annual report of the Commissioners for Relieving the Poor in Ireland show that, on the 23d of March, 1867, the number of paupers was 59,141, and 2,659 in excess of the number of the same week of 1866. The number of orphans and deserted children out at nurse in January, 1866, was 475; and 1867, at the same date, was 590.

The Rev. W. C. Van Meter, a clergyman of the Baptist Church, now missionary in Rome, was there on a tour in 1868, and, in his correspondence to America, wrote as follows:

"ROME, ITALY, *May* 20, 1868.

"Another weary day has ended, and I sit and review with a sad heart the dungeons and cells, etc., visited to-day. The very site of the inquisition makes the heart sick, by recalling the scenes of wickedness and indescribable cruelty witnessed and endured in this place. The political prisons are closed usually against all, except the one who unsuccessfully struggled for freedom, and his tormentors. The histories in it will not be revealed—except now and then a brief extract—until 'the books are opened.' I have conversed with Count ———, who was a judge sixteen years, but who arose with Garibaldi to free Italy, was captured by the Austrians, sent to one of those *hells* used by the Pope, and there endured long, sad years of cruelty and want, until, before the French troops were withdrawn, he was liberated. He made a pen out of a nail, and by some means obtained ink, and kept his journal by writing on his white shirt. I saw it. It is written all over with the sickening history of those years. Permission to visit the prisons was refused me, except in a single and unimportant case; but I did enter and examine them. How I obtained admission must not be told in this.

"The Reformatory for Boys is a miserable affair. The Foundling Hospital, for illegitimate babes, is a grand affair.

The Superior told me they have at present about four thousand, with two thousand nurses. Most of them are scattered among families. I saw one hundred and fifty to two hundred of them, and apparently well cared for. The child is placed at night in a box in the wall, and then turned into the reception-room—a bell rings, the nurse comes, and that is all that is known. I was told that every fourth child in Rome is a child of shame; for these priests and nuns, in order to induce the people to take care of them, call the child 'Figlia della Madonna'—that is, Child of the Virgin. Thus they cover the Virgin with shame, rather than with honor. They worship her; and then, as our Savior was born of her, so they call all the children not born in lawful wedlock her children. Bad as Rome is, I was not prepared for any thing so shocking.

"How I long to visit our principal cities, and tell them of what I see here, and try and stir up the people, and put them on their guard against the things that crush these people and threaten to overwhelm us."

Thus writes an American clergyman, when surrounded by the crime of Papal Rome.

Facts may be found nearer home.

Montreal, Canada, is a stronghold of Popery; and a telegram of June 4, 1868, discloses the following shocking facts:

"The report of the General Foundling Hospital of this city shows a shocking state of things. The number of children received during the last year was six hundred and fifty-two, and the number of deaths six hundred and nineteen. Of the deaths, thirty-six were under a week; three hundred and sixty-eight under a month; five hundred and eighty three under one year; six hundred and seventeen under five years,—leaving only two deaths among all the foundlings in the establishment between the ages of five and twelve. The report further shows that four hundred and twenty-four infants were received only half-clothed; eight were absolutely naked; eighteen had not even been washed, and fifteen were bleeding for want of the necessary attentions at birth; forty-six were tainted with the special disease of infamy; eight had been wounded by instruments; seven were more or less frozen, and a large number covered with vermin. One was sent from the United States in a carpetbag; another at the bottom of a basket; another in a waterbucket; two came squeezed and bruised; another strongly nailed up in a box; another with a pin stuck through the flesh. The sufferings of eight infants, as well as their chance of life, had been lessened by doses of opium. It is no wonder, therefore, that three were dead when received, twenty-eight dying, and one hundred and fifty-seven in actual disease. Most of the remainder perished through the wretched constitution inflicted on them by their parents."

This institution is in charge of the Gray Nuns, and the report was made by Drs. Laroque and Carpenter, of the Sanitary Association, and may be accepted as authentic and true.

Already American soil is polluted with these plague-spots, which offer incentives to vice, by furnishing facilities by which to screen the perpetrators. New York, Brooklyn, and Chicago can each boast of a foundling institution, and Protestants are being taxed to sustain them. The unrestrained influence of Auricular Confession, in connection with professed sacerdotal celibacy, convents, and monasteries, may soon create a necessity for them in all large American cities.

Through these institutions an immense number find early graves; and, of those who survive, without exception, they are trained as Papists, and state appropriations asked for their subsistence.

One illustration may suffice. The Legislature of New York granted a charter to a company of Romanists in

New York, in 1869, and in 1872 they made their Second Annual Report, which we have on file, in all its disgusting details; but which can not be here inserted. One or two facts may illustrate its contents. It shows that, during the preceding year, more than seven hundred foundlings died under its care, and more than nine hundred, then living, were sustained by the city at the rate of eight dollars each, and four dollars each additional by donations and contributions, making a total of twelve dollars each for more than nine hundred foundlings; or not less than $10,000 to sustain the institution the second year of its existence.

If such institutions are permitted to multiply, what will soon be the condition of our country? Are virtuous Protestant men and women willing to be taxed to encourage licentiousness, and furnish female Jesuits facilities to perpetuate a system of corruption in our midst?

The police reports of Chicago for a period of four years, from 1866 to 1870, present the fact that, when the population of the city did not probably number more than two hundred thousand, the police arrested, during that period, more than eighty-eight thousand criminals.

In the year 1868, the police of Chicago arrested twenty-two thousand and forty-three criminals, and the following table exhibits their nationalities. The figures show that twenty-one different nationalities were represented in the criminal department, and that eleven thousand eight hundred and twenty-five of them were Irish, and that more than four-fifths of them were foreigners:

NATIONALITIES.

Americans	3,084	Indians	2
Africans	787	Italians	75
Bohemians	149	Mexicans	2
Belgians	10	Norwegians	395
Canadians,	150	Polanders	2
Danes	25	Scotch	285
English	565	Swedes	208
French	229	Spaniards	3
Germans	3,096	Swiss	2
Hungarians	8	Welsh	16
Irish	11,825		
Total			22,043

Protestant Irish, Germans, and other nationalities, are noted for honest thrift and industry, and it is a notorious fact, shown by the records of crime and poverty in all the Northern cities, that the Roman Church furnishes an overwhelming majority of the criminals and paupers, for which the whole country is taxed. It also furnishes a very large proportion of the whisky sellers and drinkers. Romanists are men of like passions with ourselves, and if it were not for the corrupting theology of the Church, and the despotism and corrupting influence of the priests through the confessional they might soon rise to a more exalted position, intellectually and morally. We have startling facts which are reserved for a subsequent work.

Northern cities are cursed with Popery. Life and property are insecure, and political corruption is unparalleled, and the pillars of the nation tumble under its accumulating weight. It is clearly the result of putting corrupt men into power by a bargain and sale with the victims of Popery. Now the flood-tide is being turned into the South. The licentious monks, nuns, and Jesuits of Italy, whose property has been confiscated on account of their corrupting influence, are drifting into South America and into the Southern States of North America. And if this influence is not arrested, they will be to the South the *beginning of sorrow.*

Could we speak to be heard by every true patriot, philanthropist, and Christian in America, we would say, first and last, As you value your country, your liberty, your life, and property, and as you value the happiness of posterity, beware of the priests, monks, nuns, and Jesuits, male and female, beware of convents and the confessional!

CHAPTER XIII.

THE CONFESSIONAL A THIEF-TRAP.

UNABLE to defend Auricular Confession by Scripture and reason, Romanists have recommended it as a valuable *thief-trap,* or instrument for the restoration of stolen property.

EXAMPLE NO. 1.

"A case of conscience happened in St. Louis, which, though common enough, is always remarkable. A gentleman living in that city had a diamond necklace stolen from him some time since, which was valued at three thousand dollars, and all the ingenuity of the police had failed to ferret out the culprit. The gentleman received it lately, in perfect order, through the hands of a Catholic priest, with whom it had been left, in the confessional. Thus, in spite of what Protestants say, confession is not a bare form in our Church. The revelations made at the knee of the priest have often saved nations as well as necklaces." (Boston *Pilot*, of December 19, 1863.)

This case of conscience is reported on Roman Catholic authority, and the author admits that it is not an isolated case; that such cases are "common enough." As to the stealing part of this case, few will doubt that it is "common enough;" but as to restitution, it "is always remarkable"—so remarkable that the priest publishes it to the world. Nobody doubts that many Catholics steal, when thousands of them are annually convicted of theft by the civil authority; but the question is, What becomes of the stolen property? Not one case in a thousand has restitution to the injured person been reported.

One of two things must be true, either:

1. The priest was wrong in granting absolution when the penitent did not make restitution; or,

2. The priest was wrong in retaining the stolen property after restitution was made to him.

EXAMPLE NO. 2.

A few years since, the Rev. N. L. Rice, D. D., while residing in St. Louis, lost his *silver spoons.* Some time afterward a portion of them were returned through the confessional, and the Catholic paper of that city announced the "remarkable" fact with the usual flourish of trumpets. Dr. Rice also acknowledged the receipt of a *portion* of his spoons; but stated that *all* his spoons *had not been returned,* and modestly suggested that the priest did wrong in absolving the thief when full restitution was not made, or he did wrong in retaining the spoons after full restitution had been made to *him.*

Here are two "remarkable" cases of restitution. If the Catholic editors will refer to the records of the police and criminal courts in St. Louis, they may find a few thousand more *remarkable* cases of Romish theft, where restitution was not made, and where the stolen property, like the doctor's *spoons,* and Milly M'Pherson (the lost nun), disappeared, to be seen no more. According to Romish theology, all Catholics are *obliged* to confess all their mortal sins to the priest once each year. If they *do not* confess *all,* their confession is "void." If they do confess all (in case of theft), the priest *intentionally* permits them to retain the stolen property, or he *inten-*

tionally keeps it himself; and, in either case, he justifies the theft.

What right have priests, more than other men, to conceal stolen property? By what right do they retire into dark corners and confession-boxes, and hold *confidential* correspondence with thieves?

But we are told that "the revelations made at the knee of the priest have often saved nations as well as necklaces." Did the Jesuit priest, Garnet, attempt to prevent the *gunpowder plot*, to which he was accessary through the confessional? *No*. Did Auricular Confession save the life of Henry IV, in 1610, when the plan of his murder was known in the confessional? *No*. Did it prevent the massacre of St. Bartholomew's Day in Paris? *No*. Did it oppose the Inquisition in Spain? No; never. It was the right-arm of power to the Inquisition, as it ever has been to Papal Rome and despotic governments. Where has the Roman confessional ever saved a *Protestant* nation? Where has it ever saved the life of a Protestant sovereign? History records no such facts.

But, for the sake of an illustration, suppose we grant that the confession-box is a *thief-trap*, or "*den of thieves;*" that would not prove its utility. It only catches Catholic *thieves*, and history and facts prove that it *makes* more thieves than it *catches*.

When we were in public discussion with Priest Cogan, of Oskaloosa, Iowa, four years since, he endeavored to avoid a fair issue on Auricular Confesson, and divert attention from the theology of his Church, in the following manner:

In the midst of his speech, he seated himself on a chair on the platform, with his face to the congregation and the back of the chair at his side, and said, in substance: Ladies and gentlemen, I will now give you a specimen of Auricular Confession. I am a priest, and a priest was never known to reveal any thing from the confessional. And it is understood, of course, that there is always a screen between the priest and the penitent. We will suppose your servant-girl should come to the confessional, and I ask her if she has done any thing wrong. She says, Yes; I took a loaf of bread. I ask her how much it was worth. She says, so much. I ask if she did any thing else wrong. She says, Yes; I took more than one loaf. I ask how many. She do n't know. I ask about how long she continued to take a loaf each day. She says, About three months. I ask if she has done any thing else wrong. She says she took half a pound of butter each day, for three months. I ask if she did any thing else wrong. She says she took a ham of meat, worth so much.

Thus he endeavored to show that the confessional was a wonderful thief-trap.

In our reply, we frankly admitted that observation, and the bitter experience of many Protestants, could attest the fact that many of the most devout patrons of the confessional were expert thieves; and the more they confessed, the more they seemed disposed to steal. And one thing was certain, they were not accustomed to make restitution to Protestants. We pressed the question, to know what disposition was made of the stolen property: whether the priest, knowing the facts, sanctioned the theft by granting the thief absolution and permitting her to keep the stolen property, or whether they divided the spoils.

We referred to the fact that the criminal and police courts and prisons show that a large majority of the thieves frequented the confessional; and, while the confessional caught thieves, it only caught Roman thieves; and that it made more than it caught. That when restitution was not made to the original owner the priest ought to be held as *particeps criminis.*

Father Cogan failed to account for the stolen property; but, in the midst of his next half-hour's speech, proclaimed that he was very sick, and would not debate further at present, but would return next week and answer my lectures. It was evident that his sickness was only temporary, and not unto death, as he expected to be well next week.

On the next morning, we rode in the omnibus to the depot with Father Cogan, and he seemed so well that we congratulated him on his early restoration to health, and suggested that his recent affliction was only an attack of billious cholic from an overdose of Latin theology.

This thief-trap should receive due attention. Doubtless, millions of property is annually stolen from Protestants, for which restitution is not made.

The confessional did not save the late President of the United States from assassination. It is a fact of history, that the conspiracy was formed at the house of Mrs. Surratt, a confessing Romanist; that each of the conspirators first arrested were Romanists; that John Surratt, who fled to Italy and went into the Pope's army, was a Papist.

These facts should not be forgotten; and if justice were done, the confessors of these assassins would be held to an account for complicity in that conspiracy. They doubtless, like Priest Garnett, felt bound to conceal the facts, on account of their obligation of secrecy.

CHAPTER XIV.

THE CONFESSIONAL ENSLAVES MEN.

POPERY and despotism mutually sustain each other. The Government of the Romish Church is intensely monarchical. The priest sits as judge "in the court of conscience," and the faithful Papist, in a confessional, is arraigned as a criminal at his bar. By his Church he is taught to reverence the priest as the representative of God, clothed with plenary power to bind or to loose at pleasure, to forgive all his sins or consign him to hell forever.

The fear of the priest is ever before the eyes of the true Romanist. The will of the priest is *his will*, the word of the priest is *his law*, and obedience to the priest his paramount duty. And the true Romanist would sooner offend Almighty God than incur the displeasure of his parish priest. The man who bows his neck to the yoke of *Auricular Confession* is a *slave* the most abject. His spirit is crushed and broken; he no longer stands erect as a free man. The bondman of Egypt, under the lash of Pharaoh's taskmaster was a *free man* when compared with him. He breathes the air of liberty, treads the soil of freedom, and yet voluntarily surrenders his civil and religious liberty at the will and pleasure of a

Popish priest. *Shame* on an American citizen who will submit to such servile degradation. The priest may *compel* him to *fast* to-day and *feast* to-morrow, or he may permit him to gulp down "*rot— whisky,*" *small potatoes*, and *sour krout* to-day, and *require* him to eat fish, eggs, and *butter to-morrow*. The priest may permit him to lie, cheat, steal, murder, and break every commandment in the Decalogue, and live in communion with the Church, provided he goes to confession once each year, and once each year *eats* whole and entire the Lord Jesus Christ— "*His body, his soul, his blood, and divinity.*"

The priest may enter the domestic circle and dictate to a family of Romanists from parlor to kitchen, from the head of the family to the chamber-maid, and his will is supreme law, which they dare not violate. He may, and often does, attempt to separate husband and wife, who were lawfully married (by a Protestant minister or a civil magistrate) by pronouncing their marriage "*null and void, illicit and criminal.*"

He may, through the confessional, under false pretense, extort from widows and orphans large sums of money, or literally "*rob widows' houses, and for a pretense make long prayers.*"

If the priest imposes penance, the faithful Romanist must perform it, or otherwise incur the liability to be sentenced to endless punishment. If a Catholic comes into a Protestant church with the intention to hear and investigate the truth for himself, or if he reads a Protestant Bible, he commits a mortal sin which, if not confessed to the priest, and by him pardoned, will insure to the unfortunate subject the pains of eternal death. If such an offense against the Church was committed in a Papal country, where the power of the priesthood is unrestrained, the unfortunate subject might anticipate severe punishment, if not imprisonment and death. If the parish priest should make it a matter of conscience that his flock vote *for* or *against* a political candidate, not one would dare incur his *hot* displeasure by voting otherwise. All priests and bishops are under the dictation of the Pope of Rome, and all *orthodox* Catholics acknowledge their allegiance to him is paramount to every other obligation. The confessional is the secret sustaining and propelling power of this huge despotism.

Over the door of each confessional might be appropriately inscribed:

> " O let my strong unerring hand
> Assume thy bolts to throw,
> And deal damnation round the land
> On each I judge *my* foe."

With all our boasts of American independence, this country abounds with slaves most abject, and made so through the Romish confessional. Through the confessional the clergy enforce on the laity the observance of the intolerant dogma and ritual of the Church, and the autocratic dictum of ecclesiastical superiors. They claim the power to assess at discretion the laity for Church purposes. The Fifth Commandment of the Church makes it imperative on all to *pay* the *priest*, and a violation of this precept is a *mortal sin,* which exposes the sinner to eternal death, and the priest claims the power to enforce the penalty. Under this general system of clerical finance, the clergy have a source of immense revenue and many perquisites, among which may be enumerated pew rent, pay for masses, both for the living and the dead; fees for baptism for adults, infants, and fœtal abortions; funerals, marriages, and indulgences, etc., all of which are attended with financial considerations, and the priest has the power to enforce collections. When a Church enterprise is projected, the priest has the right to assess the laity, and under ecclesiastical penalty enforce collections. And here let it be understood, that ecclesiastical penalty in the Church of Rome involves a financial consideration. The person publicly denounced is persecuted and proscribed, and his only condition of reconciliation and absolution is implicit submission to his ecclesiastical superior. So that the Roman clergy not only profess to hold the keys of the kingdom of heaven, but by virtue of this assumption they really hold a death grasp on the purse or pocket-book of every member of the Church.

Through this means they often regulate the wages of the laboring classes, infusing a spirit of discontent, and thereby promoting if not directly instigating the mobocratic spirit which so often developes itself in popular tumults and bloodshed, mildly denominated "*strikes.*"

When railroads are being constructed, priests are detailed to collect the monthly dues from the Roman laborers, and when not promptly paid, we have knowledge of numerous instances where the priest has demanded and received it from the paymaster of the road.

The wages of servant girls are often regulated through the confessional. This may account for the fact that when the wages of one is raised in a city or place, the wages of all others are raised simultaneously, and the extra amount usually appropriated for some enterprise of the Roman Church. Thus, through the confessional, a tax is assessed on Protestants to sustain Popery, and collected from them by their hired servants, and there are living witnesses to these facts. This system of clerical espionage extends to every department of business and society.

Romanists may not lawfully marry except under clerical dictation, and their subsequent domestic intercourse as husband and wife must be regulated in the most minute details, according to the prescribed rules of their moral theology, and the priests in the confessional are required to exact detailed statements from wives relative to their strict conformity to ritual and theology in these matters. The most indelicate and disgusting questions not only may, but absolutely *must*, be answered. Considerations of modesty are not to be taken into the account.

Political matters are also subject to the same espionage and intolerant dogmatic dictation. The Roman clergy are seldom *seen* in mass meetings, or at the ballot-box; but their influence is *felt* there. Many of them never vote, and it is well that they do not. They are the subjects of a foreign despot to whom they have sworn *superior* allegiance, whose will is law, whose displeasure is perdition, and to whom they are indebted for their loaves and fishes. They have in their respective congregations fugle-men who can better subserve their interests, and who are ever present in political meetings to manipulate according to the wishes of their clerical dictators. To facilitate this work, numerous *secret societies*, under various pretenses, are organized, under the direct supervision of the Roman clergy, and often with the clergy officially at their head. Among these organizations are the Jesuits, Fenians, Hibernians, St. Patricks, Knights of St. Patrick, Father Matthew, and Father Burke societies, and others, too numerous to mention. Above all these, the so-called *"Catholic Union,"* which is the consolidation of these treasonable political secret organizations. This last named organization is the focal point around which are being concentrated the energies of the Roman clergy, and through which, aided by infidelity and the liquor influence, they hope to foist into position and power corrupt men who will subserve their sordid interests, and through whose legislation they expect to suppress the freedom of speech and of the press, degrade free schools, and thereby secure State funds for their sectarian parochial schools, and compel Protestants to contribute to support their driveling nonsense, and antiquated, disgusting mummery. Not unfrequently, in consideration of this consolidated Roman vote, pledges are exacted from unscrupulous demagogues before elections, that in their legislation, or in their official appointments they will subserve the interests of the Papacy. Thus, through the insidious influence of the confessional, virtue is prostrated, life and property jeopardized, civil and religious liberty threatened with destruction, and all to subserve the interests of Popery, whose native element is despotism, whose history is tarnished with innocent blood, and whose consecrated aggressive weapons are *racks, gibbets, fagots, dungeons, inquisitions, and infuriated mobs?* We appeal to the sons of revolutionary sires. Shall American free men yield their blood-bought, heaven-blessed institutions to the insidious intrigue and eternal hatred of a treason-working Roman hierarchy? NO! Let the emphatic response of *holy indignation* burst from the hearts of free men. NO!. Let life, property, and sacred honor be pledged; let American patriots arise and man their posts, panoplied with the weapons of justice, vigilance, and eternal truth; disseminate intelligence, protect the Sabbath, distribute the Bible, and by all laudable and peaceable means maintain evangelical piety, and thus repel the invaders, defend their institutions, and protect their liberties. And if the pall of Papal night begins to enshroud them, a foreign despot grasps at the reins of their Government, and Papal dupes, in fiendish triumph, proclaim the funeral knell of their liberties, then let American steel, wielded by freedom's potent arm, rend the Papal gloom, and the pure blood of free men wash the foul stain from the ensign of America's noble sons.

CHAPTER XV.

PROTESTANT SLAVES TO THE CONFESSIONAL.

UNWELCOME as the truth may be, many nominal Protestants are the bond slaves of Popery, and fear more the "*anathema*" of the father confessor than the wrath of Omnipotence.

There, for example, is

FARMER TRUSTWORTHY.

He has ten thousand acres, more or less, of beautiful, rich land, his house furnished, and his barns filled. He numbers his cattle by thousands, while bankers and brokers bow obsequiously to him. If there is a man on earth who may boast of personal independence, he is that man. He sincerely believes that he daily thanks God for civil and religious liberty. He once almost wept his eyes dim over the real and imaginary evils of African slavery. He would not own a slave for all the gold of California, nor be a slave for the whole world; and yet, poor, mistaken soul, he is a slave to his ignorant and superstitious hirelings, through the confessional. He dare not, at his breakfast-table, speak his sentiments on Romanism in the presence of his wife or children. But why not? The reason is obvious. There sits "Mike,"

"Pat," "Barney," "Jimmy Reagan," "Dan O'Flaherty," and a few more priest-ridden victims of the confessional, who have been taught, and who most sincerely believe, that "heretics" have no rights which Romanists are bound to respect, and that a word spoken against "Holy Mother Church" is a mortal sin deserving death and eternal perdition. The result may be easily inferred: Mike will get mad and leave, Pat will tell the priest, Jimmy and the rest will raise Satan generally, and "Farmer Trustworthy" will be left alone to hoe his corn and potatoes.

Shame on such truculent pusillanimity. He could discuss the merits of any other subject or system of religion, and, if he choose to do so, denounce any other denomination in unmeasured terms of obliquy, with their most unqualified approbation; but the mildest allusion to the errors of Romanism explodes the magazine of their concealed wrath, the effect of which he dreads more than he does the bolts of heaven's artillery. A slave most abject is he whose soul is not his own.

MERCHANT BLARNEY

Is constitutionally and habitually a pleasant, affable, easy-going gentleman, and desires the good will of mankind generally, and especially their patronage. With him the great *moral* question is, "*Will it pay?*" Enter his large store on Main Street. He will meet you with a smile, and converse fluently about prices current, Wall Street brokers, "Bulls and bears," and the general topics of the day. He seems to be a princely merchant, and a perfect gentleman at home, and possessed of a manly independence sufficient for any emergency. He professes to be a true Protestant, from principle and from choice. He becomes eloquent in the discussion of party politics, and among Protestants boldly asserts his religious principles. But introduce the subject of Romanism, and how changed his deportment! He *cautiously* glances an eye to the remote part of his store, where he has a Romanist for salesman or book-keeper, who is there to secure Roman *customers*, and perhaps at the instigation of a priest, whose interests he is endeavoring to subserve. Probably, better men have been rejected, simply because they were Protestants, and would not cater to the sycophantic wishes of the proprietor. There yet stands the proprietor, in painful suspense, looking first over one shoulder, then over the other, in breathless silence.

The reason is obvious. He has a few yards of cheap calico, worth about six cents per yard (and a hard bargain at that), and he knows that, unless he can impose it on Bridget, he will never sell it, in this world or in the next. And he must have a clear coast before he dare utter a sentence on the subject of Romanism, lest the fact should reach the priest in the confessional, and he lose the hard-earned patronage of the Church, which he has gained at the sacrifice of manly independence.

Thus the princely merchant, Blarney, with the fear

of the confessional before his eyes, is a crouching slave to his Roman customers.

DOCTOR DOOLITTLE.

A few doors around the corner is the office of Doctor Doolittle, with an old scull and a few bones lying loosely around for the edification of small children, and for the especial comfort of nervous women. A few empty quinine bottles, and many large patent-medicine advertisements, proclaim him a man of business. He looks out through his youthful glasses the very personification of wisdom, and speaks Latin so fluently that he has almost forgotten his vernacular tongue.

The subject of Romanism is introduced, and he prances around his room (six by nine), vociferating and gesticulating as if the destiny of the nation were suspended on his lips. Do you inquire what is the matter? Matter enough! He has a few sugar-coated pills, hard enough to shoot at muskrats. He knows he will never sell them on earth, unless he prescribes them for "Pat," "Mike," "Barney," or some of the rest, when they shake "*wid the ague.*" His pills must be sold if, as a consequence, the world should be damned.

With him, "one religion is about as good as another," and his moral vision never rises above his sordid interests. With him, principle is both antiquated and obsolete. He keeps one eye to the confessional, and the other to his pocket; and would sooner offend Jesus Christ than a parish priest.

GROCER DRINKALL,

The personification of brutality and vulgarity, is horrified at the thought of any presumptuous mortal daring to utter a word against the "Holy Mother Church." Its members are his best customers; and if his whisky is not strong they will take the more of it. He would be bankrupt in twelve months if all were Protestants, and true to their principles. He occasionally gives the priest a free glass for his influence in the *confessional*.

ATTORNEY ALLQUIBBLE.

He reads Blackstone, Chitty, Kent, Greenleaf, Story, etc., during the week, with a cigar in his mouth, and his heels on his office table, often higher than his head. He derives legal inspiration from the fumes of tobacco, and often intensified by bad whisky. He seldom reads the Bible; but is a theologian by intuition. He attends the Roman Church, and braces himself upon his imaginary dignity, often assuming a large amount he never possessed. He nods assent to whole pages of Latin, ten words of which he does not understand. Do you ask for the cause of his devotion? The answer is brief: he is an aspiring politician; he wants the patronage of the Church during the week, and, above all, he wants votes on election day. By name he is a Protestant, by interest a Papist. As to principle or piety, destitute of both. He bows obeisance to the priest, hoping to be remembered graciously in the *confessional*, and in a trying hour, late in the afternoon on some future election day. He is open to conviction on all great Roman questions, and willing to pledge himself in consideration of the Roman vote, that, if elected, he will advocate certain specified appropriations of public funds for Roman institutions, or that he will favor the reduction of the salaries of teachers in the public schools, and especially oppose the reading of the Bible in them. He further pledges that he will oppose, to the extent of his ability, the enactment of any law that will restrict the desecration of the Sabbath, or prohibit the sale of intoxicating liquors, whereby his Roman constituents might be deprived of Sabbath "recreation," and the indispensable privilege of selling beer and bad whisky. He further pledges, that if he can not wholly succeed in abolishing those objectionable "*sumptuary*" laws, he will endeavor to have them so modified and guarded with legal technicalities, that an expert or shyster before a packed jury will make it extremely difficult to enforce the penalty. And if there is any other scavenger work to be done, he is ready for the contract. Of course he would spurn a bribe, that, if detected, might subject him to a penalty; but, being a lawyer, he has a right to expect a *fee* for professional service. And in consideration of a few thousand dollars, pledged for the purpose of lobbying and "log rolling," he will take a contract. Under these circumstances it is not difficult for him to find assistance. Others who have been elected by similar influences, as unscrupulous as himself, may each have an "ax to grind," and they work each for the other, but all for sordid self. Thus legislation is corrupted, the people are defrauded, the country impoverished, and monopolies formed to subserve the interests of aspiring demagogues, whose election was secured through the influence of Romanists, and by the direct instigation and dictation of the confessional. A careful examination of this subject will exhibit the fact that the present corruption of this country is justly attributed to the partisan strife and the consequent promotion of unscrupulous demagogues to position and power, through the alien influence of a foreign population, many of whom have been made infidels by Popery, and others through the confessional, under the dictation of ecclesiastical superiors, or the sworn vassals of an ecclesiastical despot at Rome. In view of these and similar facts, it is not strange that honest patriots, regardless of party, are combining their influence to suppress this prolific source of political corruption.

Let the confessional be suppressed, and Roman priestcraft will lose its corrupting political power, and as a consequence, aspiring partisans will sink to obscurity in merited contempt, and their places be filled by better men.

PETER PETTIFOGGER.

This distinguished personage had the misfortune to be born a professional man, deficient in brains. His distinguishing qualifications are impudence, ignorance, and strong lungs. He is too much of a gentleman to work, and as a professional man, his ability is not appreciated.

He is an expert when whittling pine splinters and discoursing on politics in a third-rate country tavern, or at the door of a cheap store, or in a drinking saloon. He sits around loosely, waiting for something to turn up; he pries into every body's business, but, unfortunately, has but little of his own. If there is strife, or petty litigation in the community, he is sure to interest himself, and, if possible, increase the difficulties, and prevent an amicable adjustment. At primary caucusses he is ever present; he is a man of expediency. Qualification for office, and honest principle, are with him never taken into the account. At political hustings he shouts the loudest against political corruption by men in power, provided his party are out of office.

When the election approaches, he becomes officiously obsequious; he bows politely to Mike Mooney, Pat O'Flaherty, etc., and *significantly* suggests that it will be for the interest of their religion if certain candidates are elected; that he has at his disposal several hundred dollars for *free drinks,* which can be had in consideration of the votes of certain customers, and, as additional incentive, the parties to be elected are in full sympathy with their business, if not daily patrons. The motive is sufficient; Mooney and O'Flaherty are converted and pledged to the cause. Whisky flows freely, and the rabble shout for and vote for the men who paid for the whisky. And whatever may be be the result to individuals, society, or the country at large, is a matter of indifference, since a visit to the confessional, and alms for the priest, will insure absolution. Perjury and murder may have been committed in drunken debauchery, but the magic wand of Auricular Confession wipes all away, and leaves "Peter Pettifogger" and his accomplices in spotless purity to resume their works at pleasure.

Thus scavenger work is done on contract; and this bargain and sale corrupts all parties, and fills responsible offices with irresponsible men.

In party strife, Romanists are put forward by each party to control the clanish votes of their sect, and if not elected by one party, they may be elected by another, and in either case promotes them to position and power, and thereby jeopardizes the peace and safety of the people. But what does Peter care for the well-being of society? His India-rubber conscience causes him no trouble; he is willing to become all things to all sorts of men, women, and children, if, by all means, he may control a few votes on election day. He would sell his soul to the devil, and his body to the Pope, for partisan purposes.

FLORA M'FLIMSEY

Is another dupe of Papal despotism. It may be she lives in a large brown-stone mansion, thoroughly furnished. She professes to be a true Protestant, and boasts of her puritan ancestry.

Tap gently the silver bell at her door, and out comes Bridget; you are seated in the parlor, and Bridget retires. Presently Flora makes her appearance, and seems to be a lady of intelligence and independence equal to any emergency. She converses freely and frankly on the general topics of the day, not excepting metaphysics, theology, and political economy. In fact, she seems to possess personal independence sufficient to command a regiment of soldiers at Waterloo or Bunker Hill. But let the subject of Romanism be introduced, and how changed her deportment. She glances a significant eye to the sitting-room, dining-room, or kitchen, as she quietly closes the parlor door, and in a suppressed voice, almost a whisper, she announces the important fact "my servant, Bridget, is a Catholic." She could have freely criticised any other sect or party, could have censured the Congress or President of the United States; but she dare not speak her sentiments on Romanism in the presence of an ignorant servant. The reason is obvious. She knows that her servant is a spy in her house, and that a word spoken against Romanism will be reported to the priest in the confessional, and he may take away her servant, and prevent her from obtaining another, and thus compel her to wash her own spoons. Poor Flora, with all her wealth and boasted independence, she is a slave to an ignorant dupe of Popery, and she is more influenced by the muttering curses of a bachelor priest in the confessional, than by the voice of her Omnipotent Creator.

SIMON SIMPLETON.

He is a nondescript whose insignificant dimensions baffle description of pen or tongue, whose impudence is only commensurate with his ignorance. He is a forked thing, with fat meat for brains, with empty head and pocket, and, if requisite, willing to replenish both with falsehood. He has been promoted from the position of a third-rate "printer's devil," to the responsible position of a fourth-class penny-a-liner. Like a half-famished canine, he is scenting every cesspool in search of garbage. By instinct, if not by education, is a defender of the Roman faith, whisky shops, and the confessional. He is a fit subject for Popish priestcraft, and a persistent and perpetual applicant for scavenger work. In name he is Protestant, in principle nothing. He is incompetent to report an intelligent address, and in his blissful ignorance imagines himself competent to intimidate men of principle and sense, who dare defend the right. Utterly irresponsible, a slave to the confessional, so far beneath contempt that he can only be reached by the withering blast of scorn.

PARSON GOODTALK

Is pastor of a large and fashionable city Church. He belongs to a highly honored and very useful class of professional men. He is a perfect gentleman, a profound scholar in many departments, a conscientious Christian, and in his theological training he split all the fine hairs in critical exegesis between Calvin and Arminius, but, unfortunately, never sounded the bottomless gulf between the Pope of Rome and Jesus Christ. And, as a consequence, he is silent on that subject. He really knows but little of the fearful depths of corruption veiled in the "Mystery of Iniquity," and his congregation far less. Members of his Church, with their infants in their arms, and tears in their eyes, will solemnly covenant, before God and the Church, that they will train their children in the nurture and admonition of the Lord, and who, at the earliest opportunity, send them to a Roman convent or Jesuit school, to be trained in the service of the devil. The lambs of the flock are slaughtered by wolves, and the watchful shepherd gives the alarm, and, with shepherd's crook and sling, attempts to defend his flock against the ravages of the destroyer. But here he is assailed through the confessional. Not directly; this would be detected, and denounced as persecution. No; the most polished shafts of the devil's armory must be brought into requisition. There is Deacon Brown, Colonel Jones, and Alexander the Coppersmith, who are regarded as pillars of the Church, and upon whom the poor shepherd is largely dependent for pasturage and a field in which to labor. They are worldly-minded business men, engaged in heavy commercial transactions, and view all moral questions from a worldly stand-point, and decide questions of religious principle by a computation of present loss and gain. Here, again, is another opportunity to crush a faithful minister, and stab the cause of truth, through the confessional; and the priest of Rome gladly avails himself of the opportunity. Through the confessional, he hurls the weight of his congregation against the business interests of these Church officers; and they, feeling the pressure, but not comprehending its import, fly at once to the pastor for relief. They entreat him, in all sincerity, to desist; that preaching against Romanism will ruin their Church. The Church is feeble, and only able to meet expenses now, and a few more sermons on that subject will be fatal to its future success.

O, how little do such men understand the hidden mystery, Popery! How little do they comprehend the fact that they are under the instigation of a Popish priest in the confessional, muzzling the mouth and fettering the hands of a faithful minister of Jesus Christ! How little do they comprehend the fact that they are plunging a dagger to the heart of the Church they profess to love!

Such are the secret workings of Auricular Confession. It is an all-pervading spirit of intolerance; and, in all departments of society and business, to the extent of clerical power, it is used to intimidate Protestants, to prevent them from asserting their rights and defending their principles.

Special effort is made to embarrass Protestant enterprises, and intimidate public men. There is scarcely an influential political newspaper in the land, of any party, that is not held under restrictions by Popery. They are endeavoring to gain possession of many of the large public halls, and, when they can do no more, get in Romanists as agents; and in either case have power to favor Papists, and annoy or exclude Protestants. They are putting forward Romanists as reporters for papers, from whom it is next to impossible to obtain a fair and truthful report of any lecture or address by a Protestant. Special attention is given to all that relates to the interests of Popery.

Through the confessional, efforts are made to suppress the circulation of the Bible among the people, and its use in schools. Through the confessional, efforts are made to cause news-agents and even train-boys to circulate Roman literature; and in all possible ways the confessional is employed to darken and mislead the minds of its adherents. It is the secret projecting power of mobs, tumults, and assassinations. The best method of quelling a mob is to give the priest due notice that he will be held responsible for the consequences.

After careful observation for more than a quarter of a century, and in many of the strongholds of Popery in America; and after having repelled mob violence, and thwarted repeated plots of assassination, we confidently assert, as our deliberate conviction, that Roman mobs are either by direct instigation of the clergy, or the result of their intolerant theology through the confessional, or both combined. And we unhesitatingly declare that, in our opinion, the most effectual method of preventing or suppressing mob violence is, to hold the Roman clergy accountable, in person and property, for the consequences.

CHAPTER XVI.

PRISON-PENS FOR AMERICAN DAUGHTERS.

WE can not dismiss the subject of Auricular Confession in connection with convent life, without giving one more incident.

The Inquisition yet exists at Rome, and Catholic bishops from America, with the knowledge of the fact, have gone to Rome, to bow obsequiously around the Pope, and, possibly, may enjoy the exquisite pleasure of kissing his big-toe. This is their privilege, and we do not covet their pleasure. But we have a right to say something about their *prison-pens* in our own country. We have a right to look into the dungeons of high-walled convents. We have a right to demand the release of prisoners who are famishing for food. We have a right to demand the protection of the orphan, and the liberty of captives; and we appeal to Americans and Protestants to lend their influence, until the oppressed shall go free. The following is but one instance of suffering and oppression. We regard the statements as well authenticated. We are personally acquainted with some of the parties, and, from prudential reasons, withhold names.

We have other startling facts, which, if published, might jeopardize the life of the victim. Read the following letters, and, in the strength of American Protestants, resolve that the prisons shall be opened, and the oppressed shall go free:

"QUINCY, *June* 24, 1867.

"J. G. WHITE: DEAR SIR,—The inclosed letter is from a young lady who went to the convent at Belleville, one year ago, to attend school. She is an orphan; has considerable of fortune; also has Mr. ——, of this city, as guardian. She was engaged to be married when she started to the school, but wished to be better prepared to mingle in the accomplished society which her marriage would throw her in. And this, sir, is the end to which she is brought in this short time. I wish you would publish this, and perhaps it would keep some other Protestant girl from going. There are few strong enough to withstand their *power* when once under their *care*.

"QUINCY FRIEND."

"BELLEVILLE, *June* 13, 1867.

"MRS. ——: DEAR FRIEND,—You have not the slightest idea of the extraordinary pleasure that I derived from yours of May 15th. I should have written ere this, but I wished to give you a decided answer as to whether I could come home or not. With inexpressible joy would I accept your kind invitation if it were possible, but the sad news came this morning that I *can not*. O, what a smart to my heart! What a cloud hangs over my life, when I think that I shall never—no, never—more behold you, my dear and cherished friend! I had flattered myself with the vain hope that I should see you once more: vain hopes! they sadly deluded me. I shall soon part with all that is dear. I am to be received into the *Trappist* order. I will give you a slight idea of the life I shall hereafter lead. We never appear outside the walls; never smile; never speak, only when very necessary; sleep in a coffin, and each day dig a small portion of our own grave; practice all kinds of penance and fastings. Our food is bread and water chiefly, with herb soup. No flesh eaten.

"O, what a contrast! I often compare it with the past, and can hardly believe it true—sometimes imagine it a dream;

but no, it is reality. The world no longer affords me pleasure. No doubt, you will think me strange; perhaps crazy. I am not, yet. Let your thoughts be as strange as they may, they can not exceed mine. I have one request to make. I pray, I beg of you, to never efface me from your memory. O, what a consolation will it be to me, in my lonely cloister, to know that you, my dearest friend, will think of me when all others shall have forgotten me! I look upon you as my consoling angel. Oft in my solitude will I think of you. I shall never forget your dear features. No doubt, I have hitherto displeased and offended you; but I implore your forgiveness. Erelong, you may look upon me as one dead; for so I shall be to the gay and gaudy world. I believe I have written quite enough for the present.

"Please remember me to ———, and accept my love for yourself. I am, as ever, your loving

"MOLLIE."

We published the facts at the time, and appealed to the guardian of this orphan girl to rescue her, which we are informed he did immediately. Time alone can develop her future destiny. The fact that convents have their prison-pens and appliances of cruelty is no longer a matter of doubt, and that helpless females are there imprisoned for life is equally evident.

How long will slumbering Americans submit to the dictation and domination of mediæval Popery, with its dungeons and tortures? How long shall the cry of orphans be unheard? How long shall ecclesiastical brothels corrupt American youth, and defy the civil authorities to investigate their sanctimonious pretensions? If county and State prisons, alms-houses, and asylums may be inspected, why not convents and monasteries? Why should prison-pens be established in our midst, under the minions of the Pope of Rome, who yet sits brooding over the Inquisition in Rome, while patriots pine in its sanguinary vaults?

We appeal to Americans, and all friends of civil and religious liberty, to arise in your legislative sovereignty, and demand that convents shall be open to inspection, or forever closed, on American soil.

CHAPTER XVII.

PAPAL CONSPIRACY AIDED THROUGH THE CONFESSIONAL.

THE fact that Romanists are engaged in a conspiracy to destroy civil and religious liberty in America is apparent to all who have given careful attention to the subject. It is equally evident that the Papal power is broken on the Eastern Continent, and its last forlorn hope is on the Continent of America, and to this its energies are directed. Its plans are far-reaching and deeply laid, and a struggle has commenced with a desperation worthy of a doomed and despairing despotism. The spirit of Popery is inherently intolerant; it is unchanged and unchangeable for the better. The idea of a radical reformation in the system of Popery is a fallacy incompatible with history and without authority from the Bible. The system of Popery is not, was not, and never can be, an integral part of the Church of Jesus Christ. It is a huge excrescence which, amidst ignorance and superstition, has developed itself, and which has endeavored to subvert and supplant every principle of truth taught by Jesus Christ. In the whole range of Christian theology there is not a cardinal doctrine which Papal Rome has not distorted or perverted, not excepting the divinity of Jesus Christ. To a casual reader terms are often employed that seem to be in conformity to the teaching of the Bible, and are therefore more liable to mislead and deceive. The system of Popery is not to be reformed, but to be "destroyed by the brightness of His coming."

The Papal power is now, as in time past, unscrupulous as to the means to be employed, provided they accomplish the desired end. "No faith with heretics" is as true of Popery now in America as in the days of John Huss and the Inquisition. The cruel edicts of Popes and Councils are not repealed, and they are ready to be enforced if Romanists had the power. The fires of Papal persecution are not extinct, they are only obscured amidst the ruins and smoldering ashes of an ecclesiastical despotism. A favorable breeze would again fan them to a flame. The recent proclamation of infallibility is only the reassertion of the intolerant principles of the dark ages. It is worthy of Pope Gregory VII, the nefarious despot, by whom it was originated, and the corrupt and licentious Popes by whom it was maintained.

The recent proclamation of the Pope's infallibility is an insult to common sense, and a reiteration of the intolerant principles which have caused the death of millions of God's faithful servants.

The doctrine of infallibility is by many imperfectly understood. It has a direct connection with every other part of the intolerant system, not excepting Auricular Confession. Infallibility, as defined and understood at Rome and in apppoved theology, means:

1. That in "dogma" (doctrine of the Church) the Pope can not err.

2. That by Divine right he is universal spiritual sovereign throughout the world, and all are bound to obey him.

3. That by virtue of his universal spiritual supremacy he is also universal *temporal sovereign*—above king, queen, emperor, president, or constitution—and as such he requires all Romanists to swear superior allegiance to him.

Thus Doctor Dollinger and Father Hyacinthe declare that they can not be true to the government under which they live, and true to the dogma of infallibility. *Nor can a man be an orthodox Romanist and true to the Constitution of the United States.* They are as adverse as light and darkness, liberty and despotism, Christ and antichrist. No man can be true to both. Popery is confessedly an absolute monarchy. The Government of the United States is a democratic Republic. They have no affinity for each other.

Protestantism and Popery have never harmonized, nor can they ever harmonize. One or the other will, on this continent, become extinct. Papists are sanguine, and boast that they will subvert and supplant Protestantism in America. Many intelligent Protestants declare that they shall not do it. And a conflict of a fearful character is inevitable, and probably much nearer than many persons have imagined. Romanists are thoroughly organized, and many Protestants are profoundly asleep, apprehending no danger.

To arrest attention and awake a slumbering nation, let facts be exhibited, and let the people observe the fearful oaths by which Romanists bind themselves to obey the Pope.

The following is the oath taken by every Popish bishop on his consecration. It was abreviated in compliance with a request from this country, by the Pope in 1846, but nothing in sentiment or spirit was omitted:

ROMISH BISHOP'S OATH.

"I, G. N., elect of the Church of N., from henceforth will be faithful and obedient to St. Peter the Apostle, and to the holy Roman Church, and to our lord, the Lord N., Pope N., and to his successors canonically coming in. I will neither advise, consent, nor do any thing that they may lose life or member, or that their persons may be seized or hands any wise laid upon them, or any injuries offered to them, under any pretense whatsoever. The counsel which they shall intrust me withal, by themselves, their messengers, or letters, I will not knowingly reveal to any, to their prejudice. I will help them to defend and keep the Roman Papacy and the royalties of St. Peter, saving my order against all men. The legate of the apostolic see, going and coming, I will honorably treat, and help in his necessities. The rights, honors, and privileges, and authority of the holy Roman Church, of our lord the Pope and his aforesaid successors, I will endeavor to preserve, defend, increase, and advance. I will not be in any council, action, or treaty, in which shall be plotted against our said lord, and the said Roman Church, any thing to the hurt or prejudice of their persons, right, honor, state, or power; and if I shall know any such thing to be treated or agitated by any whomsoever, I will hinder it all that I can; and, as soon as I can, will signify it to our said lord, or to some other, by whom it may come to his knowledge. The rules of the Holy Fathers, the apostolic decrees, ordinances, or disposals, reservations, provisions, and mandates, I will observe with all my might, and cause to be observed by others. Heretics, schismatics, and rebels to our said lord, or his aforesaid successors, I will to the utmost of my power persecute and oppose. I will come to a council when I am called, unless I be hindered by a canonical impediment. I will by myself, in person, visit the threshold of the apostles every three years; and give an account, to our lord and his aforesaid successors, of all my pastoral office, and of all things any wise belonging to the state of my Church, to the discipline of my clergy and people, and, lastly, to the salvation of souls committed to my trust; and will, in like manner, humbly receive and diligently execute the apostolic commands. And if I be detained by a lawful impediment, I will perform all the things aforesaid by a certain messenger hereto specially empowered, a member of my chapter, or some other in ecclesiastical dignity, or else having a parsonage; or, in default of these, by a priest of the diocese; or, in default of one of the clergy (of the diocese), by some other secular or regular priest of approved integrity and religion, fully instructed in all things above mentioned. And such impediment I will make out, by lawful proofs, to be transmitted by the aforesaid messenger, to the cardinal proponent of the holy Roman Church, in the congregation of the sacred council. The possessions belonging to my table I will neither sell nor give away, nor mortgage, nor grant anew in fee, nor any wise alienate—no, not even with the consent of the chapter of my Church—without consulting the Roman Pontiff. And if I shall make any alienation, I will thereby incur the penalties contained in a certain Constitution put forth about this matter.

"So help me God and these holy Gospels of God."

A large portion of the Popish priests in this country are from Maynooth College, in Ireland. The following is the oath taken by them on being admitted to the order of priests:

ROMISH PRIEST'S OATH.

"I, A. B., do acknowledge the ecclesiastical power of his holiness and the mother Church of Rome, as the chief head and matron above all pretended Churches throughout the whole earth; and that my zeal shall be for St. Peter and his successors, as the founder of the true and ancient Catholic faith, against all heretical kings, princes, states, or powers repugnant unto the same; and although I, A. B., may follow, in case of persecution, or otherwise to be heretically despised, yet in soul and conscience I shall hold, aid, and succor the mother Church of Rome, as the true, ancient, and apostolic Church; I, A. B., further do declare not to act or control any matter or thing prejudicial unto her, to her sacred orders, doctrines, tenets, or commands, without leave of its supreme power or its authority, under her appointed, or to be appointed; and, being so permitted, then to act, and further her interests more than my own earthly good and earthly pleasure, as she and her head, his holiness, and his successors, have, or ought to have, the supremacy over all kings, princes, estates, or powers whatsoever, either to deprive them of their crowns, scepters, powers, privileges, realms, countries, or governments, or to set up others in lieu thereof, they dissenting from the mother Church and her commands."

Many Jesuits are in this country, and their number is rapidly multiplying. The following is the oath they take on joining the order:

THE JESUIT'S OATH.

"I, A. B., now in the presence of Almighty God, the blessed Virgin Mary, the blessed Michael the Archangel, the blessed St. John the Baptist, the holy apostles St. Peter and St. Paul, and all the saints and sacred host of heaven, and to you, my ghostly father, do declare from my heart, without mental reservation, that his holiness, Pope ——, is Christ's vicar-general, and is the true and only head of the catholic or universal Church throughout the earth; and that, by the virtue of the keys of binding and loosing, given to his holiness by my Savior Jesus Christ, he hath power to depose heretical kings, princes, states, commonwealths, and governments, all being illegal without his sacred confirmation, and that they may safely be destroyed: therefore, to the utmost of my power, I shall and will defend this doctrine, and his holiness's rights and customs, against all usurpers of the heretical (or Protestant) authority whatsoever; especially against the now pretended authority and Church of England, and all adherents, in regard that they and she be usurpal and heretical, opposing the sacred mother Church of Rome. I do renounce and disown any allegiance as due to any heretical king, prince, or state, named Protestants, or obedience to any of their inferior magistrates or officers. I do further declare, that the doctrine of the Church of England, the Calvinists, Huguenots, and of others of the name Protestants, to be damnable, and they themselves are damned, and to be damned, that will not forsake the same. I do further declare, that I will help, assist, and advise all or any of his holiness's agents, in any place wherever I shall be in England, Scotland, and Ireland, or in any other territory or kingdom I shall come to, and do my utmost to extirpate the heretical Protestant's doctrine, and to destroy all their pretended powers, regal or otherwise. I do further promise and declare, that notwithstanding I am dispensed with, to assume any religion heretical, for the propagating of the mother Church's interest, to keep secret and private all her agents' counsels, from time to time, as they intrust me, and not to divulge, directly or indirectly, by word, writing, or circumstance whatsoever, but to execute all that shall be proposed, given in charge, or discovered unto me, by you, my ghostly father, or any of this sacred convent. All which, I, A. B., do swear by the blessed Trinity, and blessed Sacrament, which I am now to receive, to perform, and on my part to keep inviolably; and do call all the heavenly and glorious host of heaven to witness these my real intentions to keep this my oath. In testimony hereof I take this most holy and blessed Sacrament of the Eucharist; and witness the same further with my hand and seal, in the face of this holy convent, this —— day of ——, An. Dom." etc.

OATH OF A LAYMAN.

COMMONLY CALLED THE CREED OF POPE PIUS IV.

"I, N. N., with a firm faith, believe and profess all and every one of those things which are contained in that creed which the holy Roman Church maketh use of, to-wit: I believe in one God, the Father Almighty, Maker of heaven and earth, of all things visible and invisible: and in one Lord Jesus Christ, the only-begotten Son of God, born of the Father before all ages; God of God; Light of light; true God of the true God; begotten, not made; consubstantial with the Father, by whom all things were made. Who for us men, and for our salvation, came down from heaven, and was incarnate by the Holy Ghost of the Virgin Mary, and was made man. He was crucified also for us under Pontius Pilate, suffered, and was buried. And the third day he rose again, according to the Scriptures; he ascended into heaven, sitteth at the right hand of the Father, and shall come again with glory to judge the living and the dead; of whose kingdom there shall be no end. I believe in the Holy Ghost, the Lord and the life-giver, who proceedeth from the Father and the Son: who, together with the Father and the Son, is adored and glorified; who spake by the prophets. And in one holy, Catholic, and Apostolic Church. I confess one baptism for the remission of sins; and I look for the resurrection of the dead, and the life of the world to come. Amen.

"I most steadfastly admit and embrace the apostolical and ecclesiastical Traditions, and all other observances and constitutions of the same Church.

"I also admit the Holy Scriptures, according to that sense which our holy mother the Church hath held and doth hold, to whom it belongeth to judge of the true sense and interpretation of the Scriptures; neither will I ever take and interpret them otherwise than according to the unanimous consent of the Fathers.

"I also profess that there are truly and properly Seven Sacraments of the new law, instituted by Jesus Christ our Lord, and necessary for the salvation of mankind, though not all for every one, to-wit: Baptism, Confirmation, the Eucharist, Penance, Extreme Unction, Orders, and Matrimony; and that they confer grace: and that, of these, Baptism, Confirmation, and Orders can not be repeated without sacrilege. I also receive and admit the received and approved ceremonies of the Catholic Church, used in the solemn administration of the aforesaid Sacraments.

"I embrace and receive all and every one of the things

which have been defined and declared in the holy Council of Trent, concerning original sin and justification.

"I profess, likewise, that in the Mass there is offered to God a true, proper, and propitiatory sacrifice for the living and the dead. And that in the most holy Sacrament of the Eucharist there is truly, really, and substantially the Body and Blood, together with the Soul and Divinity, of our Lord Jesus Christ; and that there is made a conversion of the whole substance of the bread into the Body, and of the whole substance of the wine into the Blood; which conversion the Catholic Church calleth Transubstantiation. I also confess that under either kind alone Christ is received whole and entire, and a true Sacrament.

"I constantly hold that there is a Purgatory, and that the souls therein detained are helped by the suffrages of the faithful.

"Likewise, that the saints reigning together with Christ are to be honored and invocated, and that they offer prayers to God for us, and that their relics are to be had in veneration.

"I most firmly assert that the images of Christ, of the Mother of God ever Virgin, and also of other saints, ought to be had and retained, and that due honor and veneration are to be given them.

"I also affirm that the power of Indulgences was left by Christ in the Church, and that the use of them is most wholesome to Christian people.

"I acknowledge the Holy, Catholic, Apostolic, Roman Church for the mother and mistress of all Churches; and I promise true obedience to the Bishop of Rome, successor of St. Peter, Prince of the Apostles, and Vicar of Jesus Christ.

"I likewise undoubtedly receive and profess all other things delivered, defined, and declared by the sacred canons and General Councils, and particularly by the holy Council of Trent. And I condemn, reject, and anathematize all things contrary thereto, and all heresies which the Church hath condemned, rejected, and anathematized.

"I, N. N., do at this present freely profess and sincerely hold this true Catholic faith, out of which no one can be saved: and I promise most constantly to retain and confess the same entire and inviolate, by God's assistance, to the end of my life."

With such fearful obligations binding their consciences, how is it possible for any true Papist to be a true and loyal citizen of this Republic?

Commencing with the blasphemous assumption of the Pope's infallibility, and consequent temporal supremacy, we have a consecutive chain of oaths, binding bishops, priests, Jesuits, and the laity, in obedience to the Pope, and the confessional to enforce the obligation, under penalty of eternal perdition if they fail to comply with the dictations of the Pope.

The confessional furnishes every facility to detect and punish the slightest delinquency on the part of any member of the Church.

This system of Popery is worse than a military despotism. The commander-in-chief of an army might court-martial and shoot to death an insubordinate officer or soldier, and there his power would end. Not so with the Pope of Rome. He claims the power, not only to put men to death, but to consign to endless perdition—all who reject his authority. Romanists believing he is possessed of such power, dare not disobey him. And if the Pope of Rome to-day were to declare the Constitution of the United States adverse to the interests of Popery, every orthodox Papist in the world would accept the declaration and unite with the Pope to destroy the Constitution. The Pope's curse of excommunication absolves subjects from their oaths of allegiance or fealty. If Pope Pius should issue an order to his bishops in America to control the Roman vote for any given sectarian purpose, or to destroy our system of free schools, every priest and layman would be under obligation to obey their ecclesiastical superiors.

If the poor old imbecile at Rome should determine to re-enact the Massacre of St. Bartholomew, the Gunpowder Conspiracy, or the scene of the invincible Armada, it would only be requisite to issue his orders, and his loyal subjects would be bound to obey him.

The hundreds of thousands of Jesuits, Fenians, Hibernians, Knights of St. Patrick, and other societies, many of whom are thoroughly armed, and under the command of the clergy, would spring to arms at the tap of a drum.

Where is the necessity for the secret, oath-bound political and military societies of Papists in our midsts under clerical dictators? Where is the necessity for armed companies of Romanists in our large cities, known nominally as "home guards?" These military organizations hold their secret meetings, which, taken in connection with public boasts and threats, are significant, and should not be lightly esteemed by American patriots. Romanists in high position have boasted of their plans and purposes.

A Roman Catholic Council which met in Baltimore a few years since, issued a circular which contains the following language:

"God has given us a work to do here in this new world, which, with boundless energy, is just springing into the full expansion of its strength and resources. The mission of Catholics is *to convert the world.* Our special and instant mission is *to convert our country!* If we do not succeed, we shall be scarcely in our graves when the deluge of impiety will sweep over the land, destroying both the Church and the State. In truth, they do not read the times nor the country aright, who dream that there is any middle course to be pursued. We must give religion to our political liberties, or our liberties, like an unregulated steam-engine, will shatter and dash in pieces, not itself alone, but us also. The United States must become a Catholic country, or it will first of all lose the vague sense of religiousness that still checks its madness; then rush into political radicalism and democratic robbery."

Brownson, the champion of the Papacy, indorsed by twenty-four bishops, says:

"The Church may be assailed—will be assailed; but we know it is founded on a rock; and the gates of hell shall not prevail against it. It is now firmly established in this country, and persecution will but cause it to thrive. Our countrymen may be grieved that it is so; but it it useless for them to kick against the decrees of Almighty God. They have had an open field and fair play for Protestantism. Here Protestantism has had free scope; has reigned without a rival, and proved what she could do, and that her best is evil; for the very good she boasts is not hers. A new day is dawning on this chosen land; a new chapter is about to open in our history, and the Church to assume her rightful position and influence. Ours shall yet become consecrated ground; and here the kingdom of God's dear Son shall be established. Our hills and valleys shall yet echo to the convent bell. No matter who writes, who declaims, who intrigues, who is alarmed, or what leagues are formed, this is to be a Catholic county; and from Maine to Georgia, from the broad Atlantic to the broader Pacific, the clean sacrifice is to be offered daily for the quick and the dead."

These words are not original with Brownson; they are but an echo of the voice of the Roman Church, and they clearly indicate the plans and purposes of that intolerant sect. Our system of religion and government is pronounced a failure, and Romanism is urged upon us as our only refuge from ruin.

The Romish Church has, at this moment, three powerful organizations at work in our midst to subvert the institutions of this country and establish Popery on the ruins thereof. THE PROPAGANDI, or society for the propagation of the Romish faith. It was organized with special reference to the establishment of Romanism in America. Its heads are at Paris, Lyons, and Venice. It has the special charge of convents and hospitals, largely controlled by female Jesuits, who are experts in proselyting, and these institutions are established for that purpose. A thorough education would thwart their purposes. Their object is a superficial literary, but a thorough Roman education. In a word, their chief business is to proselyte the daughters of wealthy or influential Protestants. A thorough literary American education is not desired. Independent, intelligent thought would blast their hopes. Special attention is given to ornamental culture, especially needle-work, embroidery, and a smattering knowledge of painting, music, French, Latin, or something to catch the eye or ear of superficial Protestants.

In brief, their whole energies are directed to proselyte Protestant daughters to Romanism. If a young lady is wealthy they urge her by all means take the veil and become a nun, and that moment all her property drops quietly into the coffers of the Church. The vows of poverty and celibacy seal her destiny, and a few years will probably consign her to a premature grave.

If she will not take the veil, then compass sea and land to marry her to a Papist, and thus control her money; but if she persists in marrying a Protestant, then rigorously apply the discipline and dogma of the Church—show her that Bishop Purcell says:

"In the first place, the marriage of a Catholic with an unbaptized person, unless a dispensation be previously obtained, is null and void and illicit and criminal."

"The subject of mixed marriages, that is the marriages of Catholics with Protestants, is one which we can not here omit, or delegate to another. It is a subject of paramount importance to the purity of the Catholic faith and the peace of families."

"The only occasion when the Catholic Church yields her reluctant consent to a mixed marriage is, when the Protestant party solemnly promises not to interfere with the faith of the Catholic party, and to suffer the offspring, that may result from the union, both male and female, to be baptized and educated in the Catholic faith."

Impress upon her mind that "she can not expect to be happy here with one from whom she must forever be separated hereafter."

Impress upon her mind the fact that Pope Pius VII, in 1808, said that "The marriage of Protestants is not valid; that their wives are concubines and their children are bastards, and that Catholics themselves are not lawfully married except in accordance with the ritual of the Church." And if all these things fail, refer her to the fact that if she marries a Protestant, she can not be married by a priest, unless she and her betrothed will first solemnly swear to have their children baptized and raised in the Roman Church, and that she must bind herself to do all she can to proselyte her husband to the Roman faith, and that if they fail to comply with these obligations they will perjure themselves.

And as authority for this obligation, refer her to "Plain Talk about Protestantism of To-day," by Patrick Donohoe, of Boston, which is being sold on both continents with the approbation of bishops. The instructions are as follows:

"MIXED MARRIAGES.

"When one party is Catholic and the other is not, the marriage is called *mixed*.

"The Church grieves at such marriages. They exhibit great indifference in matters of religion, and often entail the non-Catholic training of the offspring. For my part, I can not understand how a Christian, a Catholic, can be so forgetful of objects divine, as to choose for a companion in life a heretical woman, to be the mother of his children, the directress of his domestic life.

"The Church leaves no means untried to make us feel how repugnant these marriages are to her. She refuses them the enhancing majesty of her wedding ritual, and positively forbids her ministers to take any other part in them but that of a *witness*. Hence such marriages are contracted outside the Church, in the vestry,—no blessing, no prayer, no holy water, no surplice, no stole. Moreover, the betrothed, on both sides, must bind themselves, beforehand, and under a solemn oath, to raise in the Catholic Church *all* the children that may issue from their marriage, both boys and girls. Unless this oath is taken, the Church will not permit a mixed marriage to be contracted.

"When you then meet the child of a mixed marriage raised

in Protestantism, know that the parents have perjured themselves.

"And were even all conditions requisite for such deplorable unions fulfilled, and the matrimonial bond signed before a priest, let it be known that the Catholic party is forbidden to go before a Protestant parson. It would be a participation with heretics *in sacred* things, and a culpable allowance in favor of heresy."

"The Catholic party must also promise to do every thing, by word and example, to bring about the conversion of the non-Catholic."

"Once married in the Catholic Church, what do you need at the meeting-house? Not the matrimonial bond, for you are already joined in it. If you only go for the purpose of hearing some fine passages of the Bible relating to matrimony, it is not worth the scandal you give, and you can as well read them at home.

"Mixed marriages are a token of weakened faith. No Christian will ever stoop to such a religious incongruity, unless he be lost to all sentiments of Catholic dignity. (Pp. 147, 148.)

Thus, by this one stroke of policy, if they ever have a family, all are pledged to Popery before they are born, and the husband must come in with the rest, or be hen-pecked, through the confessional, during life.

This may be regarded as a slow process of proselyting. Be it so; it is terribly sure. If there be mothers in the future, the present daughters are prospectively the future mothers. Hence the importance of proselyting them, in order to control the rising and future generations. Ignore the fact as we may, it is nevertheless true that mothers shape the religious character of children and youth. Probably not one father in a thousand ever taught a child the Lord's Prayer.

The mother controls the infant mind. Impressions made by her are seldom effaced; hence the fearful responsibility of training the infant mind in error.

> " 'T is education forms the common mind;
> Just as twig is bent the tree's inclined."

Romanists have caught this Catalinean idea, and they are corrupting the youth, in order to control the nation. This accounts for their zeal for Roman schools and convents in this country. Nearly three-fourths of the people in Italy and Spain can neither read nor write, and other Papal countries present similar facts. Hence, this pretended zeal for education is not for a thorough literary education, but a thorough Roman education, in subordination to the sworn enemies of American liberty.

The appeals to Rome for aid to establish schools in America indicate their purposes. Bishop David, of Kentucky, in his foreign correspondence, said:

"Had I treasures at my disposal, I would multiply colleges and schools for girls and boys; I would consolidate all these establishments, by annexing to them lands or annual rents; I would build hospitals and public houses; in a word, I would compel all MY KENTUCKIANS to admire and love a religion so beneficent and generous, *and perhaps I should finish by converting them.*" (Quarterly Register, vol. 2, 1830; p. 194.)

Again, the same bishop says:

"In twenty jubilees, wherein I have presided, **more than forty Protestants** have entered the Church; a **great number** still are preparing to share the same happiness; and I have hardly gone over the half of Kentucky. (Quarterly Register, vol. 2, 1830: p. 197.)

These boasts were not made to be read by American Protestants, but, being intercepted, were translated by an American, and sent home for publication. We might multiply evidence on this subject, but for the present a few examples may suffice. The *Tablet* has an article on "Catholics and Public Schools," which may shed light on this subject, when taken in connection with the fact that every country is ignorant and degraded in proportion to the unrestrained teaching and influence of Rome.

The *Tablet* says:

"The education itself is the business of the spiritual society alone, and not of secular society. The instruction of children and youth is included in the sacrament of orders, and the State usurps the functions of the spiritual society when it turns educator. The secular is for the spiritual, is subordinated to religion, which alone has authority to instruct man in his secular duties, and fit him for the end for which his Creator has created him. The organization of the schools, their entire internal arrangement and management, the choice and regulation of studies, and the selection, appointment and dismissal of teachers, belong exclusively to the spiritual authority."

It is a significant fact that these American convents are largely patronized by Protestants, and could not be sustained without them. Hundreds of thousands of Romanists are growing up in ignorance and sin, while Romanists are compassing sea and land to entice the daughters of influential and wealthy Protestants into their proselyting schools.

Again we say, Protestants, beware of convents and Jesuit schools! They are aiding in a great conspiracy against civil and religious liberty. The Bible says "children shall rise up against their parents and cause them to be put to death;" and there is no other influence on earth so well adapted to cause them to do it as the instruction and influence of female Jesuits in convents.

The society known as the "LEOPOLD FOUNDATION" next demands attention. It is also aiding in the great conspiracy against the Government of the United States and civil and religious liberty.

It was organized in 1829, and its history is briefly given. Schlegle, a learned lecturer, delivered a series of lectures before the Austrian Cabinet, in 1828, in which he endeavored to show that Romanism and monarchical governments sympathized with and mutually sustained each other, and that Protestantism and a democratic government mutually sustained each other. He endeavored to show that the Government of the United State was the "hotbed" of European revolutions, and that it must be destroyed, or the crowned heads of Europe would fall.

Pursuant to these lectures, it is said that the Rev. Bishop Reese, at that time Bishop of Cincinnati, Ohio, went over to Austria, in 1829, and drew up the Constitution for the Leopold Society; that the Emperor of Austria became President, and the Secretary of Austria Secretary, of the society; that the Pope of Rome blessed the organization, contributed to its funds, and granted indulgences to all who contribute to it. And, since 1829, it has continued to send funds to Jesuits in America, with the avowed object of subverting the

Government of the United States, in order to establish Popery.

When we mention Jesuit, we pronounce the synonym of all villainy. There is not a law of God or man that may not, consistent with his creed, be violated with impunity, for the good of the Church. And there is not a Government in the world, of any distinction, except the United States, from which Jesuits have not been expelled, or in which they have not been suppressed, for treason against Church and State. They have recently been expelled from Italy, Spain, and Mexico, and suppressed in Germany.

They are concentrating in North America, and the Boston *Pilot* boasts that they have more Jesuits in America than in any portion of Europe, in proportion to the Roman Catholic population; and, without the spirit of inspiration, we venture an opinion that we will never have permanent peace in the United States till this treasonable organization is expelled or disbanded.

It is a singular fact that, at the present time, Romanists are making unprecedented efforts to put their patrons and friends into position and power.

It is a singular fact that the two men of the army next to the President of the United States are under Roman influence—one, if not both, held as members of the Church.

It is also a singular fact that, during the late troubles of our country, many of the officers of the army were Papists; and, when they marched through the South,

Protestant Church property was often damaged or destroyed.

There is evidently some secret Jesuit influence controlling the appointing power. And if those in authority are not apprised of this, it is on that account the more to be deprecated. Politicians and men of the world are so much engrossed with business that they lose sight of these matters, and it would be an easy matter for Jesuit influence to be imperceptibly brought to bear at Washington City, to stab to the heart the liberties of this nation. At a time like this, it certainly becomes every Christian and patriot to look well to these matters, and the ministers of Jesus Christ to lift a warning voice against the encroachments of the Papacy.

The LONDON EMIGRANT SOCIETY was organized with direct reference to the propagation of Popery in the United States. It is reported to have been organized by Roman Catholic bankers, and other men of wealth in London. Its branches extend to the different parts of Europe. They mapped the North-western States and Canada as their first field, and more recently have included the Southern States.

In their "Emigrant's Guide," which specifies their object and contains their map, we have the following territory: Ohio, Indiana, Illinois, Michigan, Wisconsin, the eastern borders of Minnesota, Iowa, and Missouri, together with Canada West. This was their first field.

The substance of their plan was:

1. To send the *surplus* population of Europe; that is,

the pauper and criminal population, who are a tax and a burden to them there.

2. To do it such a way as to create a demand for articles of British manufacture.

3. To establish Romanism in the North-west.

Every Papist aided by that society is required to obligate himself or herself to come to the parish of a priest, or in charge of a priest, and labor for three years on a bare subsistence, and, through the authorized collector, send the proceeds of their labor to the society.

Thus, American gold goes to Europe, and we get in exchange rags, infinitely worse ragged paupers, and criminals, to fill poor-houses and prisons, and tramp from door to door, and from city to city, an intolerable nuisance. States, cities, counties, towns, and individuals are taxed to support Roman Catholic thieves and paupers, systematically imported by the minions of the Pope.

And it is not enough to tax the country on account of their poverty and crime; but, to add insult to injury, the Roman clergy are using them against our system of free schools, against virtue, morality, and religion. They are the dupes of Popery, the slaves of the clergy, and the pliant victims of partisan demagogues. They are now being colonized throughout the United States, and their influence is worse than pestilence in any community. Life and property are no longer secure where they have the ascendancy.

This accounts for the fact that all Northern and Western cities are overrun with paupers and criminals, and often the laws of God and man are set at defiance by brutish mobs.

This also accounts for the fact that unscrupulous demagogues often occupy seats in legislative and congressional halls, when honest men and men qualified for the position, are utterly ignored.

This accounts for the fact that many of our large cities are governed by ignorant Papists, the Sabbath laws are trampled to the dust, and the accursed liquor-traffic is on the increase.

This is what the Duke of Richmond, a Romanist, said was the plan to destroy the liberties of this country. He was once Governor of the Canadas, and, in a speech at Montreal, he is reported as saying:

"The Government of the United States ought not to stand, and it will not stand. But it will be destroyed by subversion, and not by conquest. The plan is this: to send over the *surplus* [pauper] population of Europe. They will go over with foreign views and feelings, and will form a heterogeneous mass, and in course of time will be prepared to rise and subvert the Government."

"The Church of Rome has a design upon that country. Popery will in time be the established religion, and will aid in the destruction of that Republic. I have conversed with many of the sovereigns and princes of Europe, and they have unanimously expressed their opinion relative to the Government of the United States, and their determination to subvert it."

Judge Haliburton, a Roman Catholic gentleman, in a pamphlet, asserts that

"All America will be a Catholic country. The Roman Catholic Church bids fair to rise to importance in America.

They gain constantly. They gain more by emigration, more by natural increase in proportion to their numbers, more by intermarriages, adoptions, and conversions, than Protestants. With their exclusive views of salvation and peculiar tenets, as soon as they have a majority, this becomes a Catholic country, with a Catholic government, with the Catholic religion established by law. Is this a great change? A greater change has taken place among the British, the Medes and the Persians of Europe, the *nolumus leges mutari* people."

Again, he says, with *emphasis,* indicated by capitals:

"The co-operation of other European nations in promoting the objects of the society is most desirable, particularly those possessing a redundant population; that is, Roman Catholic, etc. The western districts may be said to have a particular claim on the patronage of France, as it was under their former sovereignty that their vast resources, and facility of connection between the Northern Lakes and the first navigable tributaries of the Mississippi, were discovered by those enterprising and amiable French Jesuit Missionaries, Hennepin and La Salle. As to Belgium and Germany, it is almost needless to call on them for greater support than is already furnished by the mass of the Catholic population daily flowing from those kingdoms into the fertile West."

Bishop England, late of Charleston, South Carolina, who, it is understood, was the Inquisitor-General of the Jesuits, on his return from Europe, in an address to his diocese, said:

"In Paris, and at Lyons, I have conversed with those excellent men who manage the affairs of the association for propagating the faith. I have also had opportunities of communication with some of the council which administers the Austrian Association. The Propaganda in Rome has this year contributed to our extraordinary expenditure, as has the holy father himself."

An editor of a Catholic journal in Europe, says, when speaking of these missions at the West:

"We must make haste—the moments are precious. America may one day become the center of civilization, and shall truth or error there establish its empire? If the Protestants are before us, it will be difficult to destroy their influence."

Their object is to gain a balance of political power, and establish Romanism by law. Mr. Brownson, the champion of Romanism, admits this fact, and after stating that the Catholic Church has a design upon this country—that it is their purpose to possess this country—that they are aided in this work by all the Catholic prelates and priests, he makes the following significant declaration:

"Heretofore we have taken our politics from one or another of the parties which divide the country, and have suffered the enemies of our religion to impose their political doctrines upon us; but it is time for us to begin to teach the country itself those moral and political doctrines which flow from the teachings of our own Church. We are at home here, wherever we may have been born; this is our country, as it is to become thoroughly Catholic, we have a deeper interest in public affairs than any other of our citizens. The sects are only for a day; the Church forever." (Brownson's Review.)

It is not by mere accident that Papists throughout the land act in concert, and vote together. And for the benefit of ill-informed and unsuspecting Protestants who imagine that Roman priests "never meddle with politics," we insert the following facts:

"In Michigan, the Bishop Richard, a Jesuit (since deceased), was several times chosen delegate to Congress from the territory, the majority of the people being Catholics. As Protestants became more numerous, the contest between the bishop and his Protestant rival was more and more close, until at length, by the increase of Protestant emigration, the latter triumphed. The bishop, in order to detect any delinquency in his flock at the polls, *had his ticket printed on colored paper.* Whether any were so mutinous as not to vote according to orders, or what penance was inflicted for disobedience, I did not learn. The fact of such a truly Jesuitical mode of espionage I have from a gentleman resident at that time in Detroit. Is not a fact like this of some importance? Does it not show that Popery, with all its speciousness, is the same here as elsewhere? It manifests, when it has the opportunity, its genuine disposition to use *spiritual* power for the promotion of its *temporal* ambition. It uses its ecclesiastical weapons to control an election.

"In Charleston, S. C., the Roman Catholic Bishop, England, is said to have boasted of the number of votes that he could control at an election. I have been informed, on authority which can not be doubted, that in New York, a priest, in a late election for city officers, stopped his congregation after mass on Sunday, and urged the electors not to vote for a particular candidate, on the ground of his being an anti-Catholic; the result was the election of the Catholic candidate." (Foreign Conspiracy, by Samuel F. B. Morse, A. M., pp. 93, 94.)

The following extract, from the pen of O. A. Brownson, the great champion of Romanism, in the *Quarterly Review,* of 1845, speaks with the approbation of the Roman Catholic bishops, as follows:

"But would you have this country under the authority of the Pope? Why not? 'But the Pope would take away our free institutions!' Nonsense. But how do you know that? From what do you infer it? After all, do you not commit a slight blunder? Are your free institutions infallible? Are they founded on divine right? This you deny. Is not the proper question for you to discuss, then, not whether the Papacy *be or be not compatible with republican government,* but whether it be 'or be not founded in *divine right?* If the Papacy be founded in divine right, it is supreme over whatever is founded only in human right, and then your institutions should be made to harmonize with it, not it with your institutions. The real question, then, is, not the compatibility or incompatibility of the Catholic Church with democratic institutions, but, is the Catholic Church the Church of God? Settle this question first. But, in point of fact, democracy is a mischievous dream, wherever the Catholic Church does not predominate to inspire the people with reverence, and to teach and accustom them to *obedience, to authority.* The first lesson for all to learn, the last that should be forgotten, is, *to obey.* You can have no government where there is no obedience; and obedience to law, as it is called, will not long be enforced where the fallibility of law is clearly seen and freely admitted. But 'it is the intention of the Pope to possess this country?' Undoubtedly. 'In this intention he is aided by the Jesuits and all the Catholic prelates and priests,' undoubtedly, if they are faithful to their religion."

"That the policy of the Church is dreaded and opposed, and must be dreaded and opposed by all Protestants, infidels, demagogues, tyrants and oppressors, is also unquestionably true. Save, then, in the discharge of our civil duties, and in the ordinary business of life, there is, *and can be, no harmony between Catholics and Protestants.*"

This language can not be misunderstood. Brownson speaks with the full indorsement of twenty-four Roman bishops in America.

The *Freeman's Journal* says:

"Irishmen learn in America to bide their time; year by year the United States and England touch each other more powerful in America. At length the propitious time will come—some accidental, sudden collision, and a Presidential campaign at hand. They will want to buy the Irish vote, and we will tell them how they can buy it in a lump, from Maine to California—by declaring war on Great Britain, and wiping off at the same time the stains of concessions and dishonor that our Websters, and men of this kind, have permitted to be heaped on the American flag by the violence of British agents."

The *Irish Journal,* of New York, says:

"For every musket given to the State armory, let *three* be purchased forthwith. Let independent companies be formed, thrice numerous as the disbanded corps—there are no arms acts here, yet—and let every 'foreigner' be drilled and trained and have his arms always ready. For you may be sure (having some experience in the matter) that those who begin by disarming you, mean to your mischief. . . . Be careful not to truckle in the smallest particular to American prejudices. Yield not a single jot of your own; for you have as good a right to your prejudices as they. Do not, by any means, suffer Gardner's Bible (the Protestant Bible) to be thrust down your throats."

Americans (and Protestants) pause and consider. Is there nothing significant in the above? What means this hatred to the Bible, this instigating national and sectarian prejudices, this purchasing muskets, this arming and drilling Irish independent companies with arms *always ready?* If there are not arms acts by which to disarm and disband such fanatics there ought to be. It is an insult to American citizens and Protestants of all denominations. And there is not the shadow of a pretext for such treasonable declarations or demonstrations. They are, in their lawful rights, protected as other citizens, and if they were not, there is a remedy without resorting to arms under the instigation of an alien clergy. It certainly is time to disband and disarm the secret Roman military organizations in our midst.

Brownson admits that hitherto they have been acting the part of fawning sycophants, truckling to political parties. These are not his words, but these are the facts, and their object has been to obtain position, power, and financial aid to build up and sustain Popery. Now they are stronger, and propose to teach the country political Romanism. He says to Romanists:

"This is our country, as it is to become thoroughly Catholic. We have a deeper interest in the public affairs than any other citizens. These sects (that is, Methodists, Baptists, Presbyterians, etc.) are only for a day; the Church (that is, the Roman sect) forever."

This language is plain and unequivocal, and ought to be sufficient to define their plans and purposes.

Another illustration of the plans and expectations of Romanists is contained in significant hints in a recent lecture of Rev. Edward M'Glynn, a Roman clergyman, at the Cooper Institute, New York. He chose, for his subject, "Our Religious Destiny," and in summing up the substance of his lecture, he said:

"This country must become Catholic, or else our religious history will not be as God designed it to be. The Catholic religion is grand enough, broad enough, noble enough, wise and prudent enough, and divine enough to bless and sanctify all the countless energies and indomitable will, the ardent affections and keen intelligence of this American nation. [Applause.] He believed for himself that the future religion of this country ought to be Catholic. Whether it will or not, is another question. This is the religion that is destined to prevail in this land. . . . All other Churches are local and national. The Catholic religion only is a unity and universal, and this nation must yearn for a religion that blesses and sanctifies the

Union, and teaches its people to labor for the preservation of the Union, and to make the Union of these States lasting and perpetual. [Applause.] It is only the Catholic religion that recommends and blesses unity, and gives additional ties to that Union which is an instinct of the American heart. The Catholic religion, therefore, ought to be the religion of the country, and, consequently, it must in the future be the religion of the country; for it is the best calculated to bless and sanctify all that is noble in it, and to bless and sanctify the glorious instinct of union which, next to the love of liberty, is the most powerful and all-controlling instinct of the American heart. (Boston *Pilot*.)

Here is a genuine specimen of Jesuit hypocrisy, and an appeal to patriots under pretext of being the friends of common liberty.

How long has it been since the Pope of Rome crushed the Italian patriots, and re-established the Inquisition in Rome? How long since the Pope advocated a dissolution of this union of States? How long since eleven out of thirteen of the Roman Catholic papers of the United States were notoriously disloyal, and the *Freeman's Journal* twice suppressed for its treasonable sentiments? How long since Roman Catholics were endeavoring to establish Romanism, by Maximilian, on the ruins of liberty in Mexico? How long since the Fenians, aided by unscrupulous demagogues, were filibustering on the Canadian border, not to promote liberty and union, but to mature plans for a permanent dissolution of the Union, and the establishment of Popery by legal enactments?

In the late troubles of our country, the Pope knew that in the North, through his abject minions, he held the balance of power; and that, if he could dissolve the Union, and annex Cuba and Mexico, he would hold the balance of power in the South. He cared nothing for the peace and harmony of this country, North or South; but he thought he saw an opportunity to promote his sordid purposes, regardless of consequences.

If the Pope and his clergy are the friends of a pure democratic republic, why do they not manifest it in Mexico, Cuba, Italy, and in other countries, where patriots are pining in prisons, or struggling in blood for liberty? Is a professedly infallible Church one thing in in Italy and another in America, the friend of liberty here and of monarchy there? Believe it who can, I can not.

In the recent Fenian raid into Canada, the plan obviously contemplated:

1. A severance of Canada from the British Crown.
2. A *temporary* confederacy.
3. Annexation to the United States.
4. By annexation and emigration, an overwhelming controlling power at the ballot-box.
5. Gradual enactments, to establish Romanism by law, and enforce its observance under penalties.
6. By this gradual process, dissolve all the bonds of national unity, and invest the Roman clergy with that ecclesiastical and temporal power which they have sacrilegiously usurped over downtrodden Italy.

It remains to be seen whether they shall fully realize their expectations. Their purposes have been thwarted, which requires a change in tactics, and a modification of their plans, so as to identify themselves more fully with the interests of partisans.

Addressing the German and Irish population, Rev. E. M'Glynn said:

"Americanism absorbs us, and the sooner we become Americanized the better. There is no use in fighting against fate. You may hold out for a while, claiming that you are Irish; but your children will be American, and will glory in the name. And the sooner the Catholic religion becomes Americanized the better. Catholic people have an extraordinary way of propagating themselves, and that is a serious question to take into consideration. The country must remain one; and, as it is extending itself in every direction, the question arises, How will the whole country be peopled? The wealth of this country is its population; and there is no religion like the Catholic for spreading its population over the earth. And that is another fact showing that the religion of the country must be Catholic."

This is plain language, and may be easily understood; but think of it—"*Catholic religion Americanized!*" A wolf in sheep's clothing! the devil transformed into an angel of light!

A new feature has recently developed itself as a part of the great Roman conspiracy. It is known as the CATHOLIC UNION.

The following "Special Dispatch" to the St. Louis *Republican,* as found in that paper, November 29, 1871, may shed light on this subject:

"*New York, November 28th.*—Thanksgiving-day will be notable in the Catholic Churches of this country for one important event, the introduction of the Catholic Union to the public. It will celebrate the day with great pomp in St. Patrick's

Cathedral, the Archbishop of New York, Dr. M'Closkey, participating, and the chancellor of the archdiocese, Father Preston, probably the most eloquent preacher of the metropolis, preaching the address. The Catholic Union is the most important confederation that the Catholics of this country have yet projected. The objects are the union of Catholics for the protection of the regents of the Catholic Church, especially those of the Pope. Its special mission will be to band together lay Catholics, and to employ them more in the service of the Church than hitherto has been done. It is expected by its friends that, when the organization of the union is completed, the Catholics of this country will attain that influence which their numbers entitle them to, but which they have not yet obtained in these States. The object of the union is not political, so far as America is concerned; but it can be readily seen that, should any such movement as the Know-nothing ever threaten the privileges of Catholic or foreign-born citizens, this body would present an unbroken front which might well awe any persons proposing unconstitutional infringement of their rights. This, however, is not the object of the organization. It will co-operate with the Catholic unions abroad in aiding the Pope in his present difficulties, according to the necessities of the hour. It is expected that each diocese in the country will form a circle. The circle in New York is already in active operation, and includes the most thoughtful and influential Catholics in the metropolis."

The manifest Jesuit proclivities of the St. Louis *Republican* give special significance to the above communication, and ought to awaken apprehension on the part of all true Protestants and patriots. The following suggestive features of the telegram are worthy of notice:

1. "The Catholic Union is the most important confederation that the Catholics of this country have yet projected."

2. "The objects are the union of Catholics for the

protection of the *regents* of the Catholic Church, especially those of the Pope."

3. "Its special mission will be to band together lay Catholics, and to employ them more in the service of the Church than has hitherto been done."

4. "It is expected by its friends that, when the organization of the union is completed, the Catholics of this country will attain that influence which their numbers entitle them to, but which they have not yet obtained in these States."

5. "It will co-operate with the Catholic unions abroad in aiding the Pope in his present difficulties, according to the necessities of the hour."

6. "It is expected that each diocese in the country will form a circle," etc.

Here we have the framework of a secret political and military Roman Catholic organization, in league with similar organizations abroad, to aid the Pope of Rome to crush the spirit of civil and religious liberty. The effort to deny that it is a political Roman Catholic organization exhibits a transparent falsehood, and the conditional threat of an "unbroken front" (at the ballot-box) discloses its true object.

The existence of such an organization, composed of Roman Catholics who are the subjects of an ecclesiastical despot, under ecclesiastical tyrants who have no common interest with the people, should awaken apprehension, and stimulate Protestants and patriots to organize for the maintenance of civil and religious liberty.

The fact that Roman Catholics have in our midst secret military organizations, with arms and ammunition, can not successfully be denied. The Fenians alone boast of three hundred thousand available men, armed and equipped for battle. They are not going to Ireland. They would accomplish nothing in Canada; and if those arms are ever used, they will be used against true patriots and Protestants in America, probably at the ballot-box.

Slumbering Americans awake! and organize on a true Protestant basis for the protection and perpetuation of civil and religious liberty. By the love you bear to God, to posterity, and your country's liberty, we again call upon you to awake and protect your country against the aggressive efforts of the enemies of civil and religious liberty; and may the God of truth and justice arm you for the contest, and crown your efforts with triumphant success.

CHAPTER XVIII.

ROMISH INTOLERANCE ENFORCED THROUGH THE CONFESSIONAL.

PROPHECY and providence indicate the present as one of the most eventful periods of the world's history. Great principles are involved, great powers are in commotion, great questions are being solved, and great results are anticipated. The nations of the world are in commotion, the rights of men are the subject of dispute, and universal liberty or protracted despotism will be the result. The questions will soon be decisively answered, whether man is, or is not, competent to govern himself; whether Bible is, or is not, the only infallible rule of faith and practice, whether the religion of Jesus Christ is, or is not, adapted to the condition of the whole world, and whether its successful propagation is to be attained by the power of God's love, or the *brute* force of man.

These adverse and conflicting principles are incorporated into, and form a constituent part of two great ecclesiastical systems which can not be harmonized. Both can not be right, both can not be true. The triumph of the one will be destructive of the other. They are known and denominated as Romanism and Protestantism.

If Romanism be true, Protestantism is heresy. It is more—it is what they say of it, "damnable heresy;" but if Protestantism be true, Romanism is a trinity of superstition, idolatry, and priestcraft. Now what are the facts. The Roman sect is the embodiment of ecclesiastical intolerance, and with its principles the monarchs and despots of earth affiliate. Protestantism is the reverse of all this. It is the living embodiment of the great principles of civil and religious liberty, predicated upon the rights of men, in conformity to the principles of justice and the word of God. Protestantism recognizes civil and religious liberty as among the dearest of man's inalienable rights. It regards them as inwrought in the constitution of man by the Creator. It permits man the use of reason, the light of revelation, and makes him socially, civilly, and morally, what he was intended to be—the arbiter of his own destiny, amenable to God and the just enactment of man, wisely instituted for the regulation of society.

Protestantism and a free democratic republic are mutually independent of each other, and yet harmonize, and each contributes to the strength of the other. Their spheres of action are entirely different, yet they are destined to stand or fall together. Man can only worship God in spirit and in truth when he expresses his unrestrained volition; and unrestrained volition can only be exercised under a free, tolerant government. It therefore becomes a matter of necessity, involving the destiny of men, that national and individual liberty be maintained, and that the Church and State exercise their powers and privileges independent of each other.

The Roman sect regards the Church as supreme, and the authorities of the State subordinate to the dictation of ecclesiastical rulers, who govern by divine right. These two systems are inherently antagonistic. A fearful and final conflict between them is inevitable. The martial hosts are gathering to the great battle, and Providence points, as with the finger of destiny, to the Western Valley as the culminating point at which the combined forces will concentrate their energies. Here, in the great West, the hottest of earth's battles is to be fought, the greatest of earth's victories to be achieved. We do not regard it as a war of words, or as a conflict of opinion only, but of great principles, involving the destiny of millions, for time and for eternity.

The conflict has commenced in words; it may end in blood; and is certainly time for Protestants to awake from their long slumber, and cast an eager eye around to discern the signs of the times. It certainly is time that Protestants should seriously consider what would be the consequences if Romanism should gain the ascendency in this country, as they boast they will. We have seen that a great Roman conspiracy is formed to destroy civil and religious liberty.

By reference to the files of the *Shepherd of the Valley*, November 1851, and published in St. Louis, with the approbation of the Archbishop, we have the following significant language:

"The Church is of necessity intolerant. Heresy she endures when and where she must; but she hates it, and directs all her energies to its destruction. If Catholics ever gain an immense numerical majority, religious freedom in this country is at an end. So our enemies say; so we believe."

"Heresy and unbelief are CRIMES; that is the whole of the matter. And in Christian countries, as in Italy and Spain, for instance, where all the people are Catholics, and where the Catholic religion is an essential part of the law of the land, they will be punished as other crimes."

These sentiments were not indorsed by a secular paper then published in St. Louis, to which the bishop's organ replied:

"Amongst our Catholic contemporaries, the *Catholic Herald* was almost alone in its strictures: others, as the *Pilot*, copied our article and indorsed what we said. The character of our journal was not called in question; and no editor, we think, has ever ventured to make our own character the subject of debate. We told the truth, and nothing but the truth; and it is not fair to sacrifice us to the prejudices of ill-instructed and timid Catholics, or of *heretics whose delicate nerves a bold statement of Catholic doctrine may happen to shock.*"

These sentiments of the bishop are fully indorsed by other distinguished Romish ecclesiastics on this continent.

According to the doctrine of the Roman Catholic Church, all Protestants are heretics, and all heretics ought to be put to death, their property confiscated and turned over to those who will put them to death, and hold it for the true Church.

This is as truly the doctrine of the Roman Catholic

Church to-day in America, as it was in Spain when the Inquisition was successfully employed to exterminate heretics. The difference is, this is yet a Protestant country, and that was a Catholic country. This permits the liberty of conscience, that did not. The intolerant doctrines of the Church are not changed in the smallest degree for the better. We have before us the "Moral Theology" of St. Liguori, published in 1846, Peter Dens, bearing date 1864, and St. Thomas, published in 1870, and which teach, in the clearest possible terms, that heretics ought to be put to death. These are secret books of the Roman clergy in our midst, and the guide of the clergy in the confessional and other duties.

Peter Dens says:

"Notorious heretics are infamous of course, and are deprived of ecclesiastical burial.

"Their temporal goods are, of course, confiscated: yet a declaratory opinion concerning the crime from the ecclesiastical judge, ought to precede the execution: because the cognizance of heresy belongs to the ecclesiastical court."

"Finally, they are deservedly visited with other penalties, even corporal, as exile, imprisonment," etc.

"Are heretics *rightly punished* with DEATH? St. Thomas answers IN THE AFFIRMATIVE. Because forgers of money, or other disturbers of the State, are justly punished with death; therefore also heretics, who are forgers of the faith, and as experience shows, greatly disturb the State. . . . This is confirmed by the command of God under the old law, that the false prophets should be killed. . . . The same is proved by the condemnation—by the fourteenth article—of John Huss in the council of Constance."

In a recent suit in court, in Kankakee City, Illinois, between Rev. Charles Chenequy and Bishop Foley, of

Chicago, these facts were brought out in a damaging way. Rev. Mr. Chenequy had been a French priest, but renounced Romanism and retained his congregation and Church property. The bishop brought suit to dispossess him and the congregation of the property. The bishop was required to testify under oath, which he did reluctantly.

With the "Moral Theology" of St. Liguori and St. Thomas in his hands, he certified that they were of the highest authority in his Church on both continents, used in their colleges and universities, and had never been repealed.

Then the bishop was requested to read in Latin, and translate into English, the following laws and fundamental principles of action against the heretics, as explained by St. Liguori and St. Thomas:

READ BY THE BISHOP.	TRANSLATED BY THE BISHOP.
"Excommunicatus privatur omni alia civili communicatione fidelium, ita ut ipse non possit cum aliis, et, si non sit toleratus, etiam alii cum ipso non possint communicare; idque in cassibus hoc versu comprehensis: Os, orare, vale, communio, mensa negatur." (St. Liguori, tom. 9, 162.)	"An excommunicated man is deprived of *all* civil communication with the faithful, in such a way that if he is not tolerated they can have no communication with him, as it is in the following verse: 'It is forbidden to kiss him, pray with him, salute him, to eat or to do any business with him.'" (St. Liguori, vol. 9, p. 162.)
"Quanquam heretici tolerandi non sunt ipso illorum demerito, usque tamen ad secundem correctionem expectandi	"Though heretics must not be tolerated because they deserve it, we must bear them till, by a second admonition,

sunt, ut ad sanam redeant ecclesiæ fidem; qui vero, post secundam correctionem, in suo errore obstinati permanent, non modo excommunicationis sententia, sed etiam sæcularibus principibus exterminandi, tradendi sunt." (St. Tomaso, tom. 4, 91.)	they may be brought back to the faith of the Church. But those who, after a second admonition, remain obstinate in their errors, must not only be excommunicated, but they must be delivered to the secular power, to be exterminated." (St. Thomas, vol. 4, p. 91.)
"Quanquam heretici revertentes, semper recipiendi sint ad pænitentiam quoties cumque relapsi fuerint; non tamen semper sunt recipiendi et restituendi ad bonorum hujus vitæ participationem . . . recipiuntur ad pænitentiam . . . non tamen ut liberentur a sententia mortis." (St. Tomaso, tom. 4, 91.)	"Though the heretics who repent must always be accepted to penance as often as they have fallen, they must not, in consequence of that, always be permitted to enjoy the benefits of this life. . . . When they fall again, they are admitted to repent . . . but the sentence of death must not be removed." (St. Thomas, vol. 4, p. 91.)
"Quum quis sententiam denuntiatur propter apostasiam excommunicatus, ipso facto, ejus subditi a dominio et juramento fidelitatis ejus liberati sunt." (St. Tomaso, tom. 4. 94.)	"When a man is excommunicated for his apostasy, it follows from that very fact that all those who are his subjects are released from the oath of allegiance by which they were bound to obey him." (St. Thomas, vol. 4, p. 94.)

The next document of the Church of Rome brought before the Court was the Act of the Council of Lateran, A. D. 1215. But as the Latin text is too long, we will give only the translation, as it was read under oath:

"We excommunicate and anathematize every heresy that exalts itself against the holy, orthodox, and Catholic faith, condemning all heretics, by whatever name they may be known; for, though their faces differ, they are tied together by their tails. Such as are condemned are to be delivered over to the existing secular powers, to receive due punishment. If laymen, their goods must be confiscated. If priests, they shall be first degraded from their respective orders, and their property applied to the use of the Church in which they have officiated. Secular powers of all ranks and degrees are to be warned, induced, and, if necessary, compelled, by ecclesiastical censures, to swear that they will exert themselves to the utmost in the defense of the faith, and extirpate all heretics denounced by the Church who shall be found in their territories. And whenever any person shall assume goverment, whether it be spiritual or temporal, he shall be bound to abide by this decree.

"If any temporal lord, after having been admonished and required by the Church, shall neglect to clear his territory of heretical depravity, the metropolitan and the bishops of the province shall unite in excommunicating him. Should he remain contumacious a whole year, the fact shall be signified to the supreme pontiff, who will declare his vassals released from their allegiance from that time, and will bestow his territory on Catholics, to be occupied by them, on the condition of exterminating the heretics and preserving the said territory in the faith.

"Catholics who shall assume the cross for the extermination of heretics shall enjoy the same indulgences, and be protected by the same privileges, as are granted to those who go to the help of the Holy Land. We decree further, that all who may have dealings with heretics, and especially such as receive, defend, or encourage them, shall be excommunicated. He shall not be eligible to any public office. He shall not be admitted as a witness. He shall neither have the power to bequeath his property by will nor to succeed to any inheritance. He shall not bring any action against any person, but any one can bring an action against him. Should he be a judge, his decision shall have no force, nor shall any cause be brought before him. Should he be an advocate, he shall not be allowed to plead. Should he be a lawyer, no instruments made by him shall be held valid, but shall be condemned, with their author."

The Roman Catholic bishop swore that these laws had never been repealed, and, of course, that they were still the laws of his Church. He had to swear that, every year, he was bound, under pain of eternal damnation, to say in the presence of God, and to read in his Breviarium (prayer-book), that "God himself had inspired" what St. Thomas had written about the manner in which the heretics should be treated by the Roman Catholics.

With an alien priesthood under a professedly infallible Pope, and with a system of intolerant theology controlling a deluded and fanatical people by promises of heaven and threats of hell, where is there national or individual security?

Let the fact be impressed deeply in the mind of every patriot, that these intolerant doctrines are now inculcated all over this land in the highest theology of the Roman Church; and we have the original Latin theology to prove it, and defiantly challenge the Roman clergy to deny or disprove the facts.

Boniface VIII is numbered in the list of popes through whom Pope Pius IX received his infallibility. He declared, in his "Unam Sanctam:"

"Uterque est in potestate ecclesiæ, spiritualis scilicet gladius et materialis. Sed is quidem *pro* ecclesia, ille vero *ab* ecclesia exercendus; ille sacerdotis, in manu regum ac militum, SED AD NUTUM ET PA-	"Either sword is in the power of the Church; that is to say, the spiritual and the material. The former is to be used *by* the Church, but the latter *for* the Church: the one in the hand of the priest, the

TENTIAM SACERDOTIS. Oportet autem gladium esse sub gladio, et temporalem auctoritatem spirituali subjici potestati. PORRO SUBESSE ROMANO PONTIFICI OMNI HUMANÆ CREATURÆ DECLARAMUS, DICIMUS, DEFINIMUS, ET PRONUNCIAMUS OMNINO ESSE DE NECESSITATE FIDEI." other in the hand of kings and soldiers, but AT THE WILL AND PLEASURE OF THE PRIEST. It is right that the temporal sword and authority be subject to the spiritual power. MOREOVER, WE DECLARE, SAY, DEFINE, AND PRONOUNCE THAT EVERY BEING SHOULD BE SUBJECT TO THE ROMAN PONTIFF, TO BE AN ARTICLE OF NECESSARY FAITH."

At least six of the highest judicial councils of the Romish Church, with the Pope at their head, have solemnly enjoined the persecution and extermination of heretics.

The duty of putting heretics to death, is among the infallible and irrevocable decrees of its General Councils, and has been indorsed by the Church as fully as the doctrines of mass, purgatory, etc.

"No computation can reach the numbers who have been put to death, in different ways, on account of their maintaining the profession of the Gospel, and opposing the corruptions of the Church of Rome. A MILLION of poor Waldenses perished in France; NINE HUNDRED THOUSAND orthodox Christians were slain in less than thirty years after the institution of the order of the Jesuits. The Duke of Alva boasted of having put to death in the Netherlands, THIRTY-SIX THOUSAND by the hand of the common executioner during the space of a few years. The Inquisition destroyed, by various tortures, ONE HUNDRED AND FIFTY THOUSAND within thirty years. These are a few specimens, and but a few, of those which history has recorded; but the total amount will never be known till the earth shall disclose her blood, and no more cover her slain." (Scott's Church History.)

"A heretic, examined and convicted by the Church, used to be delivered over to the secular power, and punished with death. Nothing has ever appeared to us more necessary. More than one hundred thousand persons perished in consequence of the heresy of Wyclif, a still greater number for that of John Huss, and it would not be possible to calculate the bloodshed caused by Luther; and it is not yet over." (Paris Univers.)

"No good government can exist without religion; and there can be no religion without an Inquisition, which is wisely designed for the promotion and protection of true faith." (Boston Pilot.)

The Pittsburg *Catholic*, alluding to the suppression of the Protestant Chapel at Rome, in 1848, says:

"For our own part, we take this opportunity of expressing our hearty delight at the suppression of the Protestant Chapel at Rome. This may be thought intolerant; but when, we would ask, did we ever profess to be tolerant of Protestantism, or favor the doctrine that Protestantism ought to be tolerated? On the contrary, we hate Protestantism; we detest it with our whole heart and soul, and we pray that our aversion to it may never decrease. We hold it meet that in the Eternal City no worship repugnant to God should be tolerated, and we are sincerely glad that the enemies of truth are no longer allowed to meet together in the capital of the Christian world."

"Protestantism of every form has not, and never can have, any rights where Catholicity is triumphant." (Brownson's Quarterly Review.)

"You ask if he [the Pope] were lord of the land, and you were in a minority, if not in numbers, yet in power, what would he do to you? That, we say, would depend on circumstances. If it would benefit the cause of Catholicism, he would tolerate you; if expedient, he would imprison you, banish you, fine you—possibly, he might even hang you; but be assured of one thing, he would never tolerate you for the sake of the 'glorious principles of civil and religious liberty.'" (Rambler.)

"The sorriest sight to us is a Catholic throwing up his cap and shouting, 'All hail, democracy.'" (Brownson's Review, October, 1851; pp. 555, 558.)

These extracts might be greatly multiplied; but, by common consent, actions speak louder than words.

It is estimated by credible historians, that, since the birth of Popery in 606, Rome has slaughtered, for the crime of heresy, by Popish persecutors, an average of more than FORTY THOUSAND of the human family for every year of its existence.

The average number of victims yearly was much greater during the dark ages, when Popery was in her glory, and reigned *despot* of the world; when, by the terrors of excommunication and interdiction, she compelled princes to butcher their heretical subjects.

Before dismissing this subject, pause, and ask yourself: Did the Prince of Peace descend to earth to establish a civil despotism? Is the God of love the author of religious intolerance? Did he who wept at the grave of Lazarus sanction the spirit of the Inquisition? Did he who prayed for his enemies, when expiring on the cross, institute the flames of *auto-da-fe?* Did he who said, "Suffer little children and forbid them not to come unto me, for of such is the kingdom of heaven," require that their sleeping dust, when bereft of the spirit, should only find a grave in the potter's-field? Did he who said, "I am the way, I am the door, I am the good shepherd, I give unto them eternal life," transfer his power and authority over the way of life and destiny of immortal souls, to murderers and assassins? Believe it who can. I could sooner believe that wolves had been divinely appointed shepherds, and hyenas the protectors of helpless infancy. I could as soon believe that devils had become intercessors, and hell had been translated to paradise.

In view of past history, the intolerant spirit of Popery, the efforts Romanists are making to subvert this Government, and their avowed plans and purposes to destroy civil and religious liberty, we again appeal to Christians and patriots to awake and prepare for the conflict.

We especially appeal to ministers, who stand as watchmen, can "ye not discern the signs of the times?" Shall the voice of prophecy and providence be unheeded? Shall the sword come and the people be not warned.

Brethren of the ministry, of all Protestant Churches, permit me to appeal to you, by the love you bear to your country, to posterity, to the souls of men, to the Bible, to the Church, and to God. Lift your warning voice against the aggressions of Popery! Rally the sacramental hosts in defense of the Bible, in defense of free-schools, in defense of virtue, in defense of liberty, and in defense of that faith once delivered to the saints; and may God Almighty help you!

Closing the volume, we address to all true patriots a

WARNING VOICE.

Romanism and Christianity are antagonistic. Between them there is, of necessity, an irrepressible conflict. This conflict is destined to be the great conflict of the nineteenth century. Prophecy and providence indicate that the present generation will be required to assume fearful responsibilities. Whatever may be the great revolutions or changes in society, they will ultimately merge into one final struggle between Truth and Error, Light and Darkness, Liberty and Despotism, Christ and Antichrist.

In America, Rome is making vigorous efforts to regain her lost power. Her plan embraces the entire Western Continent. Her chosen field for special effort is North America; her center of operation, the Northwestern States and Canada.

Her plans have special reference to emigration, education, and an aggressive effort among the Indian and colored population. Her efforts are systematically directed against the Protestant Bible, Free Schools, and a Democratic Republic. In this, Rome is aided by the Austrian and other despotic powers. A storm is gathering—dark clouds environ our horizon; the Sun of Liberty sheds a feeble ray, while many Christians and patriots seem to apprehend no danger.

The conflicts of party spirit are not the healthful concussion of jealous liberty; but the paroxysms of envy, ambition, and deadly hate. Not the breath of the zephyr, nor the gentle undulations of the lake to prevent stagnation, but the perilous commotion of powerful elements.

The stronghold of civil and religious liberty is in North America.

Organized despotism, at home and abroad, is jealous of our civil and religious liberty. The American Republic must be crushed, or the nations must be free. Protestantism must be exterminated, or Romish priestcraft will lose its power. Protestantism rocked the cradle of our liberties, defended our youth, and brought us up to noble manhood. Protestant Christianity is the guardian angel of civil and religious liberty. In it, our hope is anchored; without it, our destruction slumbers not.

God gave this country to our fathers as a *Protestant* land, in which to erect the Temple of Liberty. The Herculean work has been accomplished, and the temple stands, a monument of national glory, defying the earthquake and the tempest. Upon its towering dome, which penetrates the skies, is inscribed to its Author, in letters of light:

"Thy wisdom inspired the great institution,
  Thy strength shall support till nature expire;
And, when creation shall fall into ruin,
  Its beauty shall rise through the mist of the fire."

Let not this glorious temple be defiled by sacrilegious hands. Let not its sacred shrines be trampled by the foot of despotism. Let it never be forgotten, that "ETERNAL VIGILANCE IS THE PRICE OF LIBERTY."

Under these impressions we write; and at the risk of being traduced and persecuted by Romanists, denounced by partisan demagogues, and sneered at by pseudo-Protestants, the truth has been, and shall be, spoken in plain language, for which no apology is offered, or eulogy asked.

And while we appeal to Christians and patriots for aid and co-operation in our great work, we would say to each:

Guard well your sacred trust; transmit to posterity that civil and religious liberty which was purchased by the blood of your fathers; and when, by the Great Architect, you shall be called from labor to refreshment, let generations coming after inscribe to your memory:

> "Now, shout the praise of those
> Who triumphed o'er the foes
> Of God and Liberty!"

THE END.

www.ingramcontent.com/pod-product-compliance
Lightning Source LLC
Chambersburg PA
CBHW022016220426
43663CB00007B/1103